ORGANIC FOOD

Consumers' Choices and Farmers' Opportunities

ORGANIC FOOD

Consumers' Choices and Farmers' Opportunities

Edited by

Maurizio Canavari
Alma Mater Studiorum-Università di Bologna
Bologna, Italy

and

Kent D. Olson
University of Minnesota
St. Paul, MN, USA

# Springer

Maurizio Canavari
Dipartimento di Economia e
 Ingegneria Agrarie
Alma Mater Studiorum-Università
 di Bologna
Viale Giuseppe Fanin, 50
40127 Bologna
Italy

Kent D. Olson
Department of Applied Economics
University of Minnesota
1994 Buford Ave., 316 ClaOff
St. Paul, Minnesota 55126
USA

Library of Congress Control Number: 2006936935

ISBN-10: 0-387-39581-4
ISBN-13: 978-0-387-39581-4

e-ISBN-10: 0-387-39582-2
e-ISBN-13: 978-0-39582-1

Printed on acid-free paper.

© 2007 Springer Science+Business Media, LLC
All rights reserved. This work may not be translated or copied in whole or in part without the written permission of the publisher (Springer Science+Business Media, LLC, 233 Spring Street, New York, NY 10013, USA), except for brief excerpts in connection with reviews or scholarly analysis. Use in connection with any form of information storage and retrieval, electronic adaptation, computer software, or by similar or dissimilar methodology now known or hereafter developed is forbidden.
The use in this publication of trade names, trademarks, service marks, and similar terms, even if they are not identified as such, is not to be taken as an expression of opinion as to whether or not they are subject to proprietary rights.

9 8 7 6 5 4 3 2 1

springer.com

PREFACE

Organic Food: Consumers' Choices and Farmers' Opportunities

1. MOTIVATION FOR BOOK

Consumers' attention to food safety issues and environmental issues increased overwhelmingly in recent decades, because of their increased concern about their own health, the environment's health, and the crises and emergencies reported worldwide. Once the only option, organic agriculture has always been a production option followed by at least a few farmers all over the world. These farmers were motivated by ethical and environmental motivations, as well as by committed consumers who supported organic agriculture thanks to a separate but fairly elitist distribution channel. After a period in which the public intervention and financial support convinced an increasing number of farmers to convert into organic farming, the niche slowly became a segment in the agri-food industry.

Organic food has now become a viable alternative for an increasing number of consumers. These consumers are worried about the presence of chemicals residues and the negative consequences on the environment caused by intensive production methods. Since organic food is presently available in the common marketing channels, along with the conventional produce, consumers now have the freedom of choice between organically and conventionally produced food.

Environment-friendly production methods, such as organic agriculture, represent a way to meet society's need for a lower-impact agriculture and to cater to a category of consumers with particular preferences. Quality certification programs in both the European Union and the United States are now available to certify that food has been produced following specific, published, and accepted production methods.

Currently, a question of great relevance is whether organic farming is economically sustainable. Do the farmers need public money to maintain their profitability even after the conversion period? How many consumers are willing to pay more for buying organic food? And how much more are they willing to pay? What kind of value do they assign to organic agriculture and foodstuff?

This monograph contains several analyses that address these questions, use various methodologies, and consider a wide range of specific situations and industries in the agri-food sector, both in Italy and in the US. The focus is both on agricultural production and on retail sales. It is aimed at giving an overview of the organic sector, both in Italy and in the US, and to show how agricultural economists are performing analyses dealing with organic produce on different points in the supply chain. The book structure is thought to cope with economic issues raised by organic farming, taking into account the consumer's needs but also the managerial and budget constraints experienced by the farmers. Farm management methodologies, as well as marketing analyses have been applied to specific research topics involving several industries in the agri-food sector. The papers strive to answer questions that have a managerial relevance: e.g. are the producers ready to adopt organic farming techniques, and are the consumers willing to pay a premium price for a certified organic produce? In doing so, the book considers both the farmers' opportunities and the consumers' choices.

Most, but not all, of the contributions were presented during the 8th Padova-Minnesota Conference on "Food, Agriculture, and the Environment", held in Stout's Lodge, Red Cedar Lake, Wisconsin, on August 26–28, 2002. Starting from that base, we have added contributions that complement and expand the base and help achieve our overall goal to improve our understanding of the demand and supply conditions for organic food.

2. SECTOR OVERVIEWS

We start with two overviews of the organic sectors in our respective countries. Edi Defrancesco and Luca Rossetto discuss the trends in and development in Italy in "From Niche to Market Segment: The Growth of the Organic Business in Italy." Catherine Greene discusses the trends and developments in the United States in "An Overview of Organic Agriculture in the United States." Canavari, et al. provide a comparison of the financial profitability and viability between our two countries in "A Comparative Profitability Analysis of Organic and Conventional Farms in Emilia-Romagna and in Minnesota."

3. FARMERS' OPPORTUNITIES

Understanding the producer's perspective is necessary in order to understand the potential benefits, problems, and roadblocks that face producers and the opportunities they have in participating in growing market. Knowing their perspective also allows us to develop an understanding for potential growth and sustainablility of the supply of organic food. The four chapters in this section of the book provide a good view from four very different groups of producers: meat, crops, fruit, and wine. Based on a series of interviews with them, Luigi Galleto examines the production and market situation and the structural features and marketing strategies of organic meat producers in "Organic Meat in Italy: Situation and Perspectives in Light of the Experience of a Small Group of Firms in the Veneto Region." Mahoney *et al.*, use experimental results to analyze profitability and risk for organic and conventional crop producers in "Profitability of Organic Cropping Systems in Southwestern Minnesota." Pirazzoli, *et al.* estimate and compare the costs and profits of organic and integrated management strategies for fruit production in Comparing the Profitability of Organic and Integrated Management peach growing in Emilia-Romagna." Luca Rossetto presents an overview of the organic wine market and based on a survey of organic wine growers, discusses their farm structure, management, profitability, technology use, market, and marketing strategies in his chapter, "Marketing Strategies for Organic Wine Growers in the Veneto Region."

4. CONSUMERS' CHOICES

Consumers choose what they eat based on their own individual tastes and preferences and the supply and price of products available in the market. To have a better understanding of the market environment, we need to understand the potential level and strength of the demand for a product just as we need to understand its supply situation. In this section, we examine the demand conditions surrounding four very different products and an overview of the organic market in the U.S. Edi Defrancesco discusses the definition of standards for organic marine fish farming, production costs for organic fish farming, and the potential demand for certified organic marine fish in "Potential Demand for Organic Marine Fish in Italy." Scarpa et al. report their statistical analysis of survey data on consumers preferences for environmentally friendly production methods in "Investigating Preferences for Environment Friendly Production Practices." Alessandro Corsi and Silvia Novelli evaluate the demand for organic beef using data from a telephone survey in "Italian Consumers' Preferences and Willingness to Pay for Organic Beef." Scarpa et al. use survey data in a contingent valuation methods (CVM) to evaluate the demand for organic apples in "Willingness-to-Pay for Organically Grown Apples Among Large Retail Customers." Carolyn Dimitri and Luanne Lohr "The US Consumer Perspective on Organic Foods."

5. THE FUTURE

While these chapters and analyses provide a good picture of the organic opportinitties at this time, we know conditions change, markets become more open, and new issues develop. The organic food market is also a global market beyond Europe and the United States. To complement the analyses in the earlier sections on specific product demand and supply situations, Maurizio Canavari and Kent Olson discuss, from a broader perspective, an overall view of current events, trends, and important issues in the future in their concluding chapters, "Recent Developments and Future Issues—Italy." and "Recent Developments and Future Issues—U.S."

A limitation of the book is that even though a variety of issues and topics are taken into consideration, not every possible aspect is covered; moreover, the approaches may differ consistently according to authors' background; finally the topic is hot not only on an US-Italy perspective, but also at an EU level, and is gaining relevance also in developing countries. The aim of this book was not to offer a comprehensive view on the topic, but while its editing developed we recognized that there could be a need to do that in the future.

Maurizio Canavari and Kent D. Olson

ACKNOWLEDGEMENTS

We wish to acknowledge and thank all the people who collaborated in organizing and presenting papers during the 8th and the 9th Padova-Minnesota Conference on "Food, Agriculture, and the Environment", held in Stout's Lodge, Red Cedar Lake, Wisconsin, on August 26–28, 2002, and in Conegliano Veneto, Treviso, on August 28-September 1, 2004, for their encouragements and suggestions during the editing of this book. We also thank the other authors who contributed to this book by adding their chapters for a more complete picture of the US and Italy situation on organic food. We also acknowledge and thank Riccardo Scarpa, who had the original idea for this book, for his continued encouragement.

A special thanks to Springer's editor Susan Safren, for her assistance in the editing job and her patience in waiting the final manuscript to be ready.

This publication was partly funded and has benefited from research grants given by:

- the European Union (TH/Asia-Link/006, Contract no. 91–652)
- the Economic Research Service of the United States Department of Agriculture
- the Italian Ministry of Agriculture and Forestry Policies (DM no. 91566 29/12/2004)
- the Italian Ministry of Education, University, and Research (PRIN 2004—prot. 2004079383_002)
- the University of Minnesota
- the Alma Mater Studiorum-University of Bologna (RFO 2004, Progetti pluriennali E.F. 2004).

CONTENTS

CONTRIBUTORS..xv

SECTOR'S OVERVIEW

FROM NICHE TO MARKET: THE GROWTH OF ORGANIC BUSINESS IN ITALY 3
Edi Defrancesco and Luca Rossetto

1.	Overview on Organic Market	3
2.	The Bio-boom in Italy	5
3.	Organic Farming in the Veneto Region: An Analysis Based on the 2000 Italian Census of Agriculture Data	10
4.	References	16

AN OVERVIEW OF ORGANIC AGRICULTURE IN THE UNITED STATES.. 17
Catherine Greene

1.	Introduction	17
2.	U.S. Organic Standards and Certification	18
3.	Economic Characteristics of the U.S. Organic Agriculture Sector	21
4.	Recent State and Federal Policy Initiatives	26
5.	References	27

THE PRODUCER'S PERSPECTIVE

A COMPARATIVE PROFITABILITY ANALYSIS OF ORGANIC AND CONVENTIONAL FARMS IN EMILIA-ROMAGNA AND IN MINNESOTA.. 31
Maurizio Canavari, Rino Ghelfi, Kent D. Olson, and Sergio Rivaroli

1.	Introduction	31
2.	Objectives and Hypotheses	33
3.	Materials and Methods	33
4.	Results and Discussion	36

xi

5.	Final Remarks	44
6.	References	44

SITUATION AND PERSPECTIVES OF ORGANIC MEAT IN ITALY........... 47
Luigi Galletto

1.	Organic Meat Situation in Italy	47
2.	Observations from a Small Sample of Venetian Firms Dealing with Organic Meat	51
3.	Concluding Remarks	59
4.	References	63

PROFITABILITY OF ORGANIC CROPPING SYSTEMS IN SOUTHWESTERN MINNESOTA... 65
Paul R. Mahoney, Kent D. Olson, Paul M. Porter, David R. Huggins, Catherine A. Perillo, and R. Kent Crookston

1.	Introduction	65
2.	Background	66
3.	Study Location and Design	67
4.	Data Collection and Analysis Methods	68
5.	Results	73
6.	Conclusions	80
7.	References	81

COMPARING THE PROFITABILITY OF ORGANIC AND INTEGRATED CROP MANAGEMENT... 83
Carlo Pirazzoli, Nicola Stanzani, Alessandro Palmieri, Roberta Centonze, and Maurizio Canavari

1.	Introduction	83
2.	Materials and Methods	84
3.	Results	85
4.	Final Remarks	89
5.	References	90

MARKETING STRATEGIES FOR ORGANIC WINE GROWERS IN THE VENETO REGION.. 93
Luca Rossetto

1.	Wine from Organic Agriculture	94
2.	Overview of the Organic Wine Market	95
3.	The Organic Wine Market in Italy	96
4.	A Survey on Organic Wine Market in the Veneto Region	98
5.	Concluding Remarks	108
6.	References	111

THE CONSUMERS' PERSPECTIVE

INVESTIGATING PREFERENCES FOR ENVIRONMENT FRIENDLY PRODUCTION PRACTICES... 115
Riccardo Scarpa, Fiorenza Spalatro, and Maurizio Canavari

1.	Introduction	115
2.	Theory	117
3.	Data	119
4.	Econometric Analysis and Results	119
5.	Conclusions	122
6.	References	123

POTENTIAL DEMAND FOR ORGANIC MARINE FISH IN ITALY............ 125
Edi Defrancesco

1.	Introduction	125
2.	Methodology and Data	128
3.	Findings and Comments	132
4.	Concluding Remarks	139
5.	References	141

ITALIAN CONSUMERS' PREFERENCES AND WILLINGNESS TO PAY FOR ORGANIC BEEF... 143
Alessandro Corsi and Silvia Novelli

1.	Introduction	143
2.	Data	144
3.	Willingness to Pay for Organic Beef	146
4.	Consumers' Motivations for Not Purchasing Organic Beef	153
5.	Preferences about Selling Modalities	154
6.	Conclusions	155
7.	References	156

THE US CONSUMER PERSPECTIVE ON ORGANIC FOODS.................. 157
Carolyn Dimitri and Luanne Lohr

1.	Introduction	157
2.	US Market for Organic Food Products	158
3.	The US Organic Food Consumer	161
4.	Direct Market Sales of Organic Food Products	162
5.	The Future of the US Organic Food Market	165
6.	References	166

RECENT DEVELOPMENTS AND FUTURE ISSUES

CURRENT ISSUES IN ORGANIC FOOD: ITALY 171
Maurizio Canavari

1.	Introduction	171
2.	Farmer Issues	172
3.	Food Chain Issues	173
4.	Consumers' Issues	174
5.	Policy and Trade Issues	177
6.	Further Emerging Issues	180
7.	References	181

CURRENT ISSUES IN ORGANIC FOOD: UNITED STATES 185
Kent D. Olson

1.	Introduction	185
2.	Production Issues	186
3.	Distribution and Marketing Issues	189
4.	Policy and Trade Issues	190
5.	Future Issues	192
6.	References	192

INDEX 195

CONTRIBUTORS

Maurizio Canavari
Dipartimento di Economia e Ingegneria Agrarie, Alma Mater Studiorum-Università di Bologna
Viale Giuseppe Fanin, 50, 40127 Bologna, Italy

Roberta Centonze
Dipartimento di Economia e Ingegneria Agrarie, Alma Mater Studiorum-Università di Bologna
Viale Giuseppe Fanin, 50, 40127 Bologna, Italy

Alessandro Corsi
Dipartimento di Economia "S. Cognetti de Martiis", Università di Torino
via Po, 53, 10124 Torino, Italy

R. Kent Crookston
Department of Plant and Animal Sciences, Brigham Young University
275 WIDB, Provo, UT 84602, USA

Edi Defrancesco
Dipartimento Territorio e Sistemi Agroforestali, Università di Padova
Agripolis - Viale dell'Università, 16, 35020 Legnaro (PD), Italy

Carolyn Dimitri
U.S. Department of Agriculture (USDA) - Economic Research Service (ERS)
1800 M Street, NW, Washington DC 20036, USA

Luigi Galletto
Dipartimento Territorio e Sistemi Agroforestali, Università di Padova
Agripolis - Viale dell'Università, 16, 35020 Legnaro (PD), Italy

Rino Ghelfi
Dipartimento di Economia e Ingegneria Agrarie, Alma Mater Studiorum-Università di Bologna
Viale Giuseppe Fanin, 50, 40127 Bologna, Italy

Catherine Greene
USDA ERS, Room S4051
1800 M Street NW, Washington DC, 20036

David R. Huggins
USDA-ARS, Land Management and Water Conservation Research
Department of Crop and Soil Sciences, Washington State University
215 Johnson Hall, Pullman, WA 99164, USA

Luanne Lohr
Department of Agricultural and Applied Economics, University of Georgia
314 Conner Hall, Athens, GA 30602, USA

Paul R. Mahoney
AgCountry FCS
1900 44 Street S, Fargo, ND 58103, USA

Silvia Novelli
Centro Studi per lo Sviluppo Rurale della Collina, Università diTorino
Via G. Testa, 89, 14100 Asti, Italy

Kent D. Olson
Department of Applied Economics, University of Minnesota,
1994 Buford Ave., 316 ClaOff, St. Paul, MN 55126, USA

Alessandro Palmieri
Dipartimento di Economia e Ingegneria Agrarie, Alma Mater Studiorum-Università di Bologna
Viale Giuseppe Fanin, 50, 40127 Bologna, Italy

Catherine A. Perillo
Department of Crop and Soil Sciences, Washington State University
231 Johnson Hall, Pullman, WA 99164, USA

Carlo Pirazzoli
Dipartimento di Economia e Ingegneria Agrarie, Alma Mater Studiorum-Università di Bologna
Viale Giuseppe Fanin, 50, 40127 Bologna, Italy

CONTRIBUTORS

Paul M. Porter
Department of Agronomy and Plant Genetics, University of Minnesota
411 Borlaug Hall, 1991 Buford Circle, St. Paul, MN 55108, USA

Sergio Rivaroli
Dipartimento di Economia e Ingegneria Agrarie, Alma Mater Studiorum-Università di Bologna
Viale Giuseppe Fanin, 50, 40127 Bologna, Italy

Luca Rossetto
Dipartimento Territorio e Sistemi Agroforestali, Università di Padova
Agripolis - Viale dell'Università, 16, 35020 Legnaro (PD), Italy

Riccardo Scarpa
Department of Economics, Waikato Management School, University of Waikato
Hillcrest Road, Hamilton, New Zealand

Fiorenza Spalatro
Formerly at Dipartimento di Economia Politica, Università di Siena
Piazza S. Francesco 7, 53100 Siena, Italy

Nicola Stanzani
CRPV-Centro Ricerche Produzioni Vegetali
Via Vicinale Monticino, 1969, 47020, Diegaro di Cesena (FC), Italy

ns** **SECTOR'S OVERVIEW**

FROM NICHE TO MARKET: THE GROWTH OF ORGANIC BUSINESS IN ITALY

Edi Defrancesco and Luca Rossetto [*]

SUMMARY

In last decade the organic agriculture has steadily increased, especially in EU countries where many scandals have boosted organic food consumption. Italy is still the first EU country as organic land and farms but Italian organic consumption is lower than other EU members. Actually, the Italian organic agriculture growth has been accomplished by EU grants which have moved organic market out of its niche size. More recently, supermarket chains have also shown a growing interest for organic market, but export still accounts for a great share of Italian organic production. The growth of organic agriculture has been investigated at the micro-level showing organic farm structure and organization features such as size, type of production, technology employed, and farmers characteristics such as age, education, etc., using Census data. Results about organic agriculture in the Veneto Region show that competitiveness of organic farms is strictly associated with EU grants.

1. OVERVIEW ON ORGANIC MARKET

The organic agriculture has steadily expanded world-wide in the last years not only in Europe, Japan or North America but also in many developing countries. The international market of organic food has rapidly increased as well. Organic food's share of the total food market is still small (around 1–3%) but its potential growth in the near future is enormous.

---

[*] Edi Defrancesco is full professor in Agricultural Economics at Dept. TeSAF, University of Padova, e-mail: edi.defrancesco@unipd.it. Luca Rossetto is associate professor in Agricultural Economics at Dept. TeSAF, University of Padova, e-mail: luca.rossetto@unipd.it. Sections 1 and 2 are based on Luca Rossetto. Section 3 is based on Edi Defrancesco.

Organic agriculture emerged in Europe in 1924 when Rudolph Steiner held a course in biodynamic agriculture. During the 1930s and 1940s important research was carried out by Hans Mueller in Switzerland, Lady Eve Balfour and Albert Howard in Britain, Masanobu Fukuoka in Japan. In Europe, some conventional farms converted their activity to organic farming in the 1960s.

Since the 1990s, the development of organic agriculture in European Union (EU) countries has been supported by financial subsidies. In other countries outside EU, the organic agriculture growth has been prompted by a growing demand for organic produce in Europe, United States and Japan.

According to the SÖL survey (Yussefi and Willer, 2004), the world organic area is around 24 millions hectares (MHa) mostly located in Australia (10 MHa), Argentina (2.9 MHa), Italy (1.1 MHa), USA (0.9 MHa), Brazil (0.8 MHa), Uruguay, and UK (0.7 MHa each), Germany and Spain (0.6 MHa each), France and Canada (0.5 MHa each). A greater share of the total organic area is extensive grazing land, especially in Australia and Argentina where extensive livestock systems are very common. Organic farming involves 11% of total area in Austria, 10% in Switzerland, 8% in Italy, 7% in Finland, around 6% in Denmark and Sweden, 4% in Germany, but it is between 2 and 1% in many countries such as Australia, Argentina and France, and below 1% in USA and Brazil.

In 2002, the market for organic food was estimated around 23 billion dollars (ITC, 2002). The largest organic market is the U.S. where sales have expanded to 13 billion USD in 2003, equal to 1–2% of total food sales (Shaota, 2004).

The second largest organic market is Europe where sales have reached 10.5 billion USD in 2002 (Shaota, 2004). The main European market is Germany (3.06 billion USD, equal to 2.3 % of total food sales), followed by UK (1.5 billion USD; 1.0%), France and Italy (1.3 billion USD each; 1 and 1.5%), Switzerland (700 million USD; 2.1%), Austria (330 million USD; 2.5%). In Asia the most important organic market is still Japan where sales, around 350 million USD in 2003, have decreased 10-fold since 2001 due to restrictive Japanese regulations on organic farming and organic food (ITC, 2001; FAS, 2004).

The medium term growth remains high in the U.S. (15–20%) while decreasing or stable in many in many EU countries (8%). The growth in Europe has been boosted by many scandals such as dioxin chicken, BSE, foot and mouth disease, etc.. Furthermore, this growth has been accomplished by many supermarket chains entering into the organic market, while increasing product availability and promoting consumption.

An important factor behind the organic success is the positive consumer awareness on health and environmental issues, including the resistance towards GMO farming and genetically modified food produces. Recently, increasingly aggressive and targeted marketing by the retail sector have introduced organic products and pushed sales, especially in U.S. (ITC, 2001; Gardner, 2000).

Notwithstanding this successful growth, several risk factors have to be considered. First, the fast growing market may lead to oversupply in some products. The potential for reduced price premiums for organic products and lower profitability may discourage organic farming. Second, the lack of state regulations for organic agriculture may hamper trade, especially in developing countries.

While official regulations exist in many countries, they differ in content and effectiveness (USDA-ERS, 2002). Within the EU, regulations on organic agriculture have been set by regulation 2092/91 since 1992.

Throughout the world, the International Federation of Organic Agriculture Movements (IFOAM), established the IFOAM Accreditation Program (IAP) in 1992 to provide equivalence among official regulations existing in many countries. The IAP is based on international IFOAM standards, which are developed continually, and includes rules for the use of the "IFOAM accredited" logo (IFOAM, 2002). FAO and WHO developed their official guidelines on organic food products aiming at consumer protection and information, and promoting trade (FAO, 2001). These guidelines for the production, processing, labeling and marketing of organically produced food are collected in the Codex Alimentarius that are in line with EU-regulation 2092/91 and the IFOAM basic standards. Accordingly, the National Organic Program (NOP) has fixed standards for US organic agriculture (USDA, 2000).

2. THE BIO-BOOM IN ITALY

In 2002, the total value of Italian organic production reached 1.4 million EUR, corresponding to 1.5% of total food consumption (ISMEA, 2004). Organic production does not match final consumption because production includes organic feed and exports (about 1/3 of total production). Also, a significant share of organic product is sold without certification as well (1/10 of total production). A yearly survey[1], shows that almost 2/3 of organic retail sales include four food categories: dairy (25.4%), biscuits (13.9%), fruit & vegetables (13.8%) and beverages (10.3%) (Table 1). The share of organic consumption is around 1.24% but it varies greatly among food categories reaching the highest value in case of eggs, baby and light food where consumers pay great attention to food security.

Table 1. Organic food sales trend in Italy by category (million EUR)

	2001	2002	2003	% of organic consum.	% of total consum.
Fruit & vegetables	29,895	40,052	40,629	13.8	2.15
Pasta and rice	13,537	15,188	14,411	4.9	0.73
Bread & cereals	3,010	4,323	4,222	1.4	0.80
Oils (vegetable)	6,913	10,739	10,596	3.6	1.12
Milk and dairy	57,503	76,728	74,967	25.4	1.34
Biscuits, sweets, snack	39,627	42,600	40,947	13.9	1.03
Beverages	24,108	25,860	30,453	10.3	1.18
Eggs	14,426	18,413	19,979	6.8	7.57
Dressing	6,578	9,611	12,726	4.3	1.44
Light food	9,277	9,505	4,665	1.6	5.05
Baby food	20,248	17,849	17,478	5.9	5.54
Sugar, tea, coffee	7,024	9,956	10,379	3.5	0.76
Ice cream & frozen food	10,483	10,480	10,019	3.4	0.54
Other bio food	1,957	2,700	3,553	1.2	0.24
TOTAL	**244,586**	**294,004**	**295,024**	**100.0**	**1.24**

Source: ISMEA/AC-Nielsen, 2004.

---

[1] ISMEA, an Italian research Institute, investigates organic consumption together with AC-Nielsen using a sample of consumers buying products having a EAN code label. This sample does not include organic produce sold as loose or without any EAN code label.

In 2002, the average organic expenditure per consumer[2] was around 24 USD, per year while the corresponding growth rate between 1999 and 2002 was about 30% per year (Yussefi and Willer, 2004; ISMEA, 2002).

Lately, organic market expansion has been supported by the continuous crises of conventional agriculture and animal husbandry, and it has been accomplished especially by the large-scale retail (LSR) promotion. In the 1990s, most Italian supermarket chains (Coop, Esselunga, Giesse, Pam, etc,) launched their own private label and at the beginning of 2000, the number of supermarkets with an organic corner was greater than the number of specialized organic shops. Actually, the organic market has been mainly driven by supermarket chains since 2000 (Lunati, 2002). The introduction of private labels by supermarket chains has improved grading and services in supplying organic food products. Actually, the growth in supermarkets follows the overall growth of the organic market. The growth in supermarkets can also be considered as a new market segment as opposed to only a niche market.

Now, other than supermarkets and specialized independent organic food shops (smaller than 100 square meters), there are large outlets (between 200 and 500 square meters) and about fifty regional and nation-wide franchise shops (Compagnoni *et al.*, 2002; Santucci and Pignataro, 2002). The most important is the franchisor NaturaSì with 33 franchisee "*superettes*", followed by Bottega and Natura (15 sell point). Most organic retailers are located in Northern Italy with more industrialization and higher incomes than the central and southern regions. Another relevant marketing channel is catering. There are about a hundred organic restaurants (mostly vegetarian and macrobiotic ones), most of these are located in the northern and central regions and in larger towns.

A very interesting marketing channel is the organic school cafeterias, where an organic menu is served to more than 380.000 children in nursery and middle schools located mostly in metropolitan areas (Rome, Bologna, Turin, Padua) but also in smaller towns. Starting in 1999, an Italian regulation forced municipalities and hospitals to use organic food in their catering services[3].

Almost 55% of organic retail sales is driven by supermarket chains and 31% by specialized shops (ISMEA, 2004). About 9% of organic food is sold directly by producers and 3% is sent to catering services. Two percent is used by bakeries or sold as meat in butcher shops.

Premium prices for organic products are also relevant, especially at retailing for processed imported goods. On average, the price of organic produces is 36% higher than conventional food (Table 2).

---

[2] The largest per-capita spending on organic food is found in Switzerland (USD105), followed by Denmark and Sweden with about 71 USD, while the European average expenditure is about USD 27. In non-European countries, such as the US, the average expenditure is about USD 63.

[3] Even if the law is compulsory, there is no punishment in case of failure to observe, so if parents or organic farmers do not actively advocate for it, the organic catering may not happen.

Table 2. Comparison of prices of organic and conventional food produces in different retail shops

Produce	Organic shops	Direct marketing	Supermarket, organic	Supermarket, not-organic
Olive oil	8.4	10.3	5.8	5.2
Potatoes	1.7	1.4	1.4	1.0
Tomatoes	1.8	2.8	2.3	1.5
Onions	2.0	2.0	1.8	0.9
Cucumber	2.8	1.8	1.5	2.1
Carrots	1.9	1.5	1.8	1.3
Apples	2.3	1.8	2.7	1.9
Oranges	1.5	1.4	1.6	1.2
White wine	3.8	2.1	4.1	3.4
Yogurt	4.3	n.a.	4.5	3.7
Eggs	0.3	0.3	0.4	0.2
Baby food	7.0	n.a.	8.4	6.8

Price EUR/kg or EUR/liter or EUR per egg.
Source: Compagnoni et al., 2002.

Due to small-scale processing plants and inefficiencies in distribution channels, processed food, even from domestic producers, often has very large premium prices.

However, supermarket chains are able to reduce premium levels of minimizing logistic and distribution costs since they sell both organic and conventional products.

Almost 1/3 of Italian organic production is exported to EU countries mainly, but also to the U.S. and Asian countries such as Japan or Taiwan (Gallas, 2000). The main export market is definitively Germany, whose share is about 50% of total Italian exports, followed by UK (16%) and Switzerland (14%). The residual export is sold in Austria, Scandinavian countries, France and non-EU countries. The range of organic product exported is wide: fruits and vegetables, extra virgin olive oil, wine, cheeses, sauces, condiments, delicatessen foods, pasta, cereals and pulses, dried-fruit, ice-cream and industrial products.

Consumers buying organic products are mostly located in the northern regions of Italy, where the industrial and economic structure is stronger. Recently, a survey[4] showed that 73% of Italians give a correct definition of organic and know some key characteristics (no chemicals, more naturalness). Twenty-two percent give vague definitions ("healthy, genuine, safer"). Thirty-eight percent of Italians have bought an organic product at least once. Only one out of six was disappointed, and the others (23% of all adults) were regular consumers. This last percentage increases to 48% if potential consumers who have the intention to buy organic products are added. The survey also shows that the average consumer of organic products is between 30 and 45 years old, lives in a city or large town in the north of the country, has an average or higher than average education, and is in the upper middle or upper income bracket.

The key success factors in increasing organic consumption are price, availability and information (ISMEA, 2004). In particular, the price of organic products is often higher

---

[4] This research was carried out in May 2001 by Demoskopea, a leading Italian market research institute.

than conventional prices, especially for medium-low income consumers. The availability of organic retail stores is restricted to main cities, especially in Northern Italy, with small towns and southern regions excluded. Information is sometimes lacking or biased and the market is ambiguous since consumer cannot recognize organic from conventional produce. Therefore, controls on organic systems are often not clearly defined or enforced.

The tremendous growth of the Italian organic market can be described as a bio-boom for Italian agriculture.

In December 2003, the Italian organic sector accounted for 42,185 farms with about one million hectares of land (Figure 1); 1,849 farms with processing plants; 4,264 processing and trade companies; and 175 importers. The organic area was about 8% of the total agricultural land, while the number of organic farms was 2.2% of the total number. At the EU15 level, Italian organic agriculture is almost 24.4% of the EU organic land and more than 40% of the EU organic farms (Lunati, 2004).

Regionally, 64% of organic farms are located in Southern Italy, 14% in Central Italy and 22% in the North. However, 48% of trading companies and processors and 82% of importers are found in the northern regions. At the end of 2003, 53% of organically managed landwas forage and pasture; 21% grains and cereals; 18% fruit, olives, and wine; and 4% vegetables and industrial crops. There is also a relatively large organic area cultivated to organic spices and herbs (4%).

From 1999 to 2001, the average yearly growth of Italian organic agriculture has been around 7% for farms and 9% for area. Conversely, in the last two years organic agriculture has decreased because of shortage in the EU subsidies supporting organic farming. In the 1990's, the organic agriculture growth has not been homogeneous in Italy for neither the rate of farmers entering organic systems nor land use. In the second half of the 1990s the bulk of producers, as well as, land entering the organic market (i.e., land shifting from conventional to organic systems), were mostly located in southern regions (Figure 2). Actually, most of this land was devoted to grazing as a result of farmers applying for EU grants to formally convert their pastures and meadows from the conventional to the organic system. After introducing organic animal husbandry (EC regulation 1804/99), many farmers having organic pastures also applied as organic breeders. The growth in organic livestock has been remarkable in extensive productions such as sheep, goats and horses, and in fast growing animals such as chickens (Table 3). Conversely, the shifting from conventional to organic has been rather slow in intensive productions such as beef, veal and pig since breeders find organic breeding much more restrictive than the conventional method.

The EU agro-environmental programs have had a strong influence on organic agricultural growth: positive from 1992 to 2001 and negative after 2001 because of diminishing EU funds for organic agriculture. This is not entirely conclusive, however. In Italy both new adopters and existing organic farmers are eligible for grants, but at the national level only 50% of certified farms appears to have taken advantage of the EU's Organic Aid Scheme. Some farmers are more interested in the premium prices paid by the organic market than in a grant. In other cases, transition costs (time for filling in forms, transportation costs to go into the regional government offices, etc.) are considered higher than the value of the grant itself. Nevertheless, the boom in organic farms and agricultural land during the late 1990's, especially in southern regions, has been broadly driven by the EU financial support (EC regulation 2078/92). From 2001, this financial support

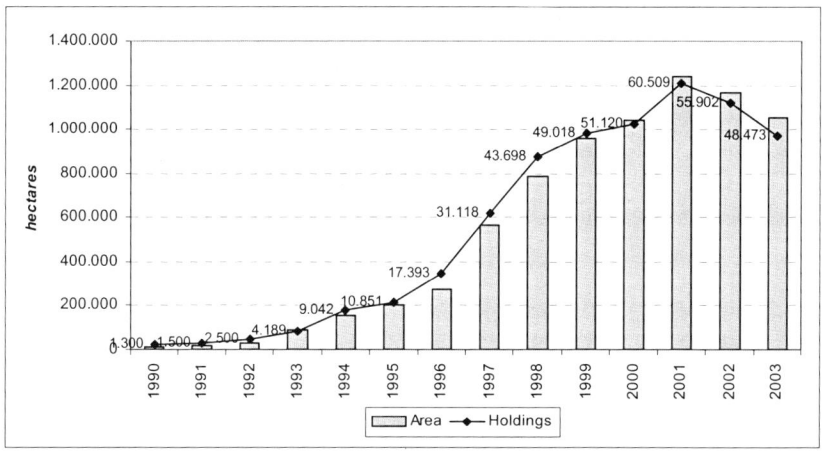

Figure 1. Development of organic agriculture in Italy 1990–2003
(Source: Lunati, 2004 on Mipaf and certification agencies data)

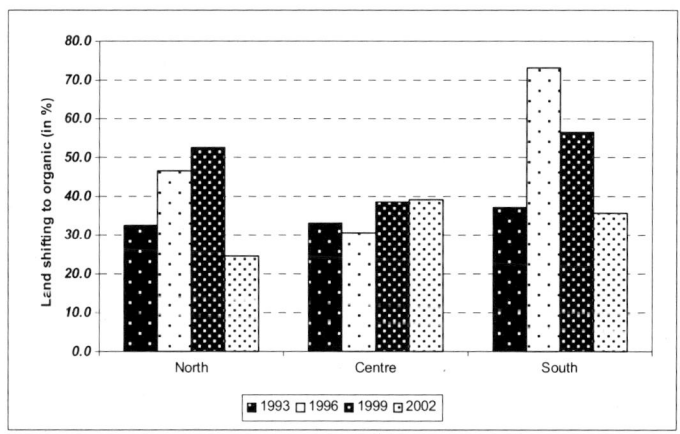

Figure 2. Land shifting from conventional to organic system per geographical area
(Source: ISMEA, 2004 on MIPAF data)

coming from agro-environmental programs provided by Agenda 2000 (EC regulation 1257/99), has been cut deeply.

Historically the Italian pioneering experiences in organic agriculture date back to the 1960's, but it only took off in the 1970's, when more and more farmers and consumers sought an improved quality of life and consumption. Once EC Regulation 2092/91 was implemented, many small associations of organic farmers, producers and consumers

Table 3. Organic animal husbandry in Italy (2002)

	Head	% change 2002/01	% organic on total
Horses	3,333	*51.2*	6.6
Beef & veal	164,536	*−50.2*	2.6
Sheep & Goat	668,451	*103.9*	8.6
Pigs	19,917	*−19.5*	0.2
Poultry	939,396	*44.8*	0.5
Bees (n. beehive)	1,377	*39.7*	n.a.

Source: ISMEA 2004 on MIPAF and ISTAT data.

reorganized themselves and joined forces through mergers and federative networks. Today, there are sixteen officially recognized certification agencies operating in Italy.

According to the most important Italian certification agency, a small share of organic farms ask for certification of their production. In 2000, AIAB controlled 13,607 farms but only 3,000 asked for certification of their product. Consequently, it can be assumed that only a small share of organic farms sells their produce as organic (Pinton, 2001). The reasons can be explained by the following factors:

- Many farms are still in the conversion period. Farms are waiting to be allowed to sale their produces as organic.
- Most organic land is pasture or cultivated to fodder crops located mostly in southern regions while dairy farms are mainly found in Northern Italy[5];
- Many farmers pay little attention to product certification because they are already satisfied by EU compensations;
- Farmers sell their (often small) organic production through a marketing channel where personal interrelationships matter more than certification labels.

Yet, the critical mass supplied by individual farms is often low, and organic products cannot be sold directly to supermarkets or shops, Generally, the main outlet market is the wholesaler. Direct sales to consumers or other shorter ways remain explored by only a few producers. Foreign markets are still unknown to most farmers.

3. ORGANIC FARMING IN THE VENETO REGION: AN ANALYSIS BASED ON THE 2000 ITALIAN CENSUS OF AGRICULTURE DATA

At the present time, Veneto is one of the most important regions in Italy for organic food production. In 2003 it accounted for:

- 1,261 organic farms managing about 20 thousand hectares (2.9% of total Italian organic farms, compared to 1.9% in 2000);
- 416 organic agro-food processing holdings (9.8% of Italian share); and
- 175 organic food importers (16% of total Italian importers, compared to 11.9% in 2000)[6].

---

[5] For example, in Sardinia there are 499,000 hectares of organic pastureland while only 8 dairies are recognized.
[6] Italian Ministry of Agriculture database on organic sector (SINAB).

For the first time in Italy, the 2000 Italian Census of Agriculture provides specific information on organic farms. In particular, Census data cover all existing organic farms: both those operating under the control of an authorized Certifying and Control Body ("certified organic") and those directly selling their produce to a limited number of consumers thanks to the farm's reputation with no reason to require external certification and pay the corresponding costs. The Census also provides detailed information not only on organic land use (forage crops excluded) and animal husbandry, but also on farm structure, e.g. land ownership, type of farm organization, family and paid labor, and the farmer's socio-economic characteristics. The organic crop acreage in the Census questionnaire is reported with no distinction between organically managed land under a conversion regime from a previous conventional management, and land producing organic produce according to the EU Regulations (EC Reg. 2092/91 and subsequent modifications).

In order to properly understand the Census data analysis summarized in this section, it should also be noted that the number of farms harvesting organic crops has been steadily increasing over the last decade, thanks to the 1991 EU Regulation and the EU agro-environmental crop payments assured to organic farmers. Organic crop farming is substantially stable at present, while organic animal husbandry can be considered in the early stages, because of its more recent EC Regulation, dated 1999.

The 2000 Census of Agriculture reports 995 farms harvesting organic crops (0.5% of total Veneto farms) and 340 practicing organic animal husbandry (0.2% of farms in the region). Excluding forage crops[7], 4,981 hectares are planted with organic crops (0.6% of regional utilizable agricultural area (UAA). Cereals (23.5% of total organically farmed land), vineyards (20.1%) and fruits (19.9%) dominate, with a similar distribution to that of conventional land use. The ratio of area planted with an organically grown crop against the total area used for that crop (organic and conventional management) in Veneto varies from 3.6% for fruits, around 2% for vegetables and olives, 1.4% for vineyards, to 0.3% for cereals. Organic livestock rearing accounts for 0.7% of total regional head of cattle, 1% of pigs, 0.6% of poultry and over 15% of sheep.

Organic farms play a more important role in the regional agriculture scenario in terms of labor. The total number of people involved in organic farming (4,356) against total agriculture workers equals to 1.2%, a share twice that of organic land use. In particular, organic farm labor accounts for 0.9% of total regional family workers (including the farmer) and a relevant share of total paid farm workers (6.4% of total full-time workers and 4.8% of temporary workers).

Organic crops are mainly concentrated in areas specializing in high quality produce and where the demand for organic produce is relatively high, e.g. the province of Verona (29% of the regional organic land) and the province of Venezia (26%).

A joint analysis of farms producing differentiated products (i.e., organic, obtained using environmentally-friendly production methods, with an EU Protected Designation of Origin (PDO) or Protected Geographical Indication (PGI) label), highlights a similar geographical distribution, a common (and above average) tendency to establish a formal and/or informal network with other farmers and a similar farm structure in terms of labor, land and capital (Defrancesco, 2003). Focusing on organic farms, the most relevant

---

[7] According to other sources on Italian organic farming, not entirely comparable with Census data, forage crop acreage was over 2,500 hectares in 2000.

factors differentiating them from all regional farms reported in the 2000 Census of Agriculture can be summarized as follows:

- *Larger farm size.* Taking into account the farm as a whole, the average size per holding of organic farms is above the general mean of all farms operating in the Veneto region: i) total mean UAA per farm is 12 hectares (5 ha of which are organically farmed) *versus* a total regional UAA average of 4.5 ha; ii) mean number of herd in the livestock inventory is higher for both ruminants and monogastric animals, excluding pigs. On the other hand, organic farms also demonstrate the traditional, fragmented structure of regional agriculture, with around 60% of organic holdings having less than 5 cultivated hectares.
- *Significantly different owned/rented land ratio.* The higher average farm size of organic holdings is partly due to a higher quota of land rented from others: the per farm mean share of rented land against total cultivated area is 40% in organic farms, whereas the average ratio at a regional level is 16%. This structural adjustment is probably due to an average expected higher profitability of organic products than conventional products (due both to market price and EU agro-environmental subsidies). This allows organic farmers to sustain the fixed costs for rented land.
- *Higher average number of labor units per farm and a more efficient ratio of workers to annual work units[8].* Due to a larger average farm size and more labor intensive production methods, organic farms report: i) an above average mean number of total workers per farm, both in terms of family workforce and external labor; ii) a significantly higher percentage of non-family paid workers; iii) a lower level of hidden underemployment by comparing the total workers/annual working units ratio to the percentage of paid temporary workers (Table 4).
- *Lower part-time farming.* Higher added unit value of organic produce and the more intensive labor required encourage full-time farming. The 2000 Census reports a higher percentage of full-time farmers (Table 5) and a lower rate of part-time among family workers (Table 4) on organic farms than on total farms of the Veneto region. External occupation of part-time organic farmers is more frequently connected to the agricultural sector than is general. It is also interesting that organic farms are generally less dependent on external machinery hiring contractors: 52% of organic farms use only their own farm machinery, whereas this percentage is 37% for all Veneto farms.
- *Younger farmer.* Principal organic farm operator's mean age (51 years) is significantly lower than on conventional farms (59 years). Over 25% of farmers managing farms growing organic crops and around 20% of organic animal husbandry operators are under 40 years old, while only 11% of total regional farmers are of the same age (Table 5). On the other hand, the frequency of over-60s is significantly lower in organic farms.

---

[8] An Annual Work Unit (AWU) corresponds to a conventional full-time worker (working 1800 hours per year). A part-time worker is converted into a pro rata AWU.

Table 4. Labor characteristics by farming type

	Organic crop production	Organic animal husbandry	Total farms Veneto
Average number of total workers per farm (*)	3.4	2.9	1.9
Average number of total workers per farm (%)			
Family workers (farmer included)	65.0	78.3	92.3
Paid full-time workers	7.5	15.7	1.7
Paid temporary workers	27.5	6.0	6.0
Total	100.0	100.0	100.0
Average number of days per hectare of utilized agricultural land	123.0	118.0	96.0
Annual working units/Total workers	0.6	0.7	0.3
Percentage of family workers with an external occupation (excluding farmer)	26.7	24.3	28.2

* family workers (farmer included) and paid workers.

Table 5. Farm manager's characteristics by farming type (percentages)

		Organic crop production	Organic animal husbandry	Total farms Veneto
Gender:	male	78.4	82.1	77.8
	female	21.6	17.9	22.2
Age:	Less than 30 years old	5.1	3.5	2.1
	30–40	20.3	15.6	8.6
	40–50	26.0	30.0	16.2
	50–60	24.1	20.6	23.9
	60–80	22.9	29.4	44.8
	over 80	1.5	0.9	4.4
Educ.:	Univ. level: agric. degree	2.4	0.9	0.4
	Univ. level: Non agric. degree	3.9	1.2	1.6
	Secondary school: agriculture	11.4	10.9	2.8
	Secondary school: other	22.5	15.0	11.1
	Intermediate level	28.2	30.0	23.1
	Primary school	30.2	38.2	56.6
	none	1.4	3.8	4.4
Seminars and courses during job		37.1	34.7	9.7
Off-farm occupation: On-farm, full time		77.7	81.7	75.4
Part-time farming		22.3	18.3	24.6
Off-farm occupation linked to agriculture (% of total part-time farmers)		17.8	15.0	8.9

- *Higher education level.* The number of farmers with a university-level education is higher on organic farms (Table 5). More generally, organic farmers have more often received a specific education in agriculture, at a secondary school and/or a university, plus seminars and training courses during their working life. In particular, the expressed demand for general, technical and economic training is particularly high (over 30% of organic farmers versus around 10% in general).

- *Relatively lower percentage of individual or family holders.* It is interesting to note that a higher percentage of businesses controlled by more than one person or stockholder is reported for organic farms than in the general case (Table 6).
- *Diversification of farm activities.* First of all, a greater diffusion of quality differentiated products (e.g. PDO-PGI labeled products) is observed among organic farms than in the general case. On the other hand, the 2000 Census highlights a significantly higher frequency of organic holdings with earning activities other than agricultural production, e.g. tourism activities, contractual work for the protection, and the improvement of the landscape and the environment under EU Agro-environmental Regulations, processing of farm produce (Table 7).
- *Use of modern technology for accessing the market.* As expected, "traditional" direct farm sales to end-consumers are particularly relevant in organic farms. For example, the percentage of direct sales ranges from 28% to 43% for fresh vegetables, being 44% for potatoes, 19% for fruits and 50% for poultry. Indeed, the rate of diffusion of both farm WEB pages informing consumers about farm activities and produce and of e-commerce among organic farms is higher than that observed at the general level (Table 8).
- *More integration among farms* in order to better access the input and output markets. In particular, Table 9 shows the greater diffusion of horizontal links among organic farms based on formal relationships. A more general level of vertical integration is also observed in regional organic farms, mainly based on production or sale contracts of organic produce.

In conclusion, the 2000 Census of Agriculture points out the structural and organizational characteristics of organic farms in the Veneto region, which differentiates them from conventional farms, at least on average terms. These characteristics seem to offer better opportunities for organic farms to operate in more competitive markets. However, it should be noted that organic farming development is still dependent on public support. This includes financial support through such methods as: i) EU partially decoupled grants to organic farms on a per hectare/per head base; ii) public incentives in the case of investments or promotion activities; iii) the relevant number of organic farms receiving an EU payment linked to agro-environmental services (Table 7) and through low-cost access to frequent training programs aimed at improving farm human capital. Nevertheless, the recent decline observed in the number of organic holdings at the Italian level (–14.3% in 2003 with respect to the previous year), compared with the 5% reduction accounted for in the Veneto region, suggests that the regional organic sector is more capable of operating under conditions of both reduced public financial support and a limited consumer growth due to the stagnating economy.

Table 6. Legal structure of the farm by type of farming (percentages)

	Organic crop production	Organic animal husbandry	Total farms Veneto
Sole proprietor	90.0	91.8	96.3
Partnership	6.7	4.7	2.7
Corporation	2.3	2.6	0.5
Cooperative	0.4	0.6	0.3
Other	0.6	0.3	0.1
Total	100.0	100.0	100.0

Table 7. Percentage of holdings with earnings other than for agricultural production by type of farming

	Organic crop production	Organic animal husbandry	Total farms Veneto
Agritourism	4.8	6.8	0.4
Other recreational services	1.7	3.5	0.2
Processing of farm produce	22.4	28.8	12.7
Handcrafts	0.2	0.3	0.0
Wood processing	0.4	0.3	0.1
Contractual work (agro-environmental services)	1.9	4.1	0.2

Table 8. Percentage of holdings accessing Internet and e-commerce by farming type

	Organic crop production	Organic animal husbandry	Total farms Veneto
Internet access	5.2	3.5	0.4
Farm WEB page	3.7	2.9	0.3
e-commerce to sell produce	1.3	0.9	0.1
e-commerce to buy inputs	1.3	0.9	0.1

Table 9. Percentage of holdings establishing a formal horizontal integration with other farms by farming type

	Organic crop production	Organic animal husbandry	Total farms Veneto
Farms' Consortium	2.6	2.9	0.7
Cooperatives	65.0	50.0	17.4
Producers' Association	37.6	54.1	7.3

4. REFERENCES

Compagnoni, A., Pinton, R., and Zanoli, R., 2002, *Organic Farming in Italy*, Stiftung Ökologie & Landbau (SÖL), Bad Dürkheim Germany; http://www.organic-europe.net.

Defrancesco, E., 2003, I cambiamenti dell'agricoltura veneta visti attraverso i dati censuari. Il sistema delle produzioni di qualità: verso un sistema integrato regionale?, in: *Rapporto 2003 sul sistema agroalimentare del Veneto*, Veneto Agricoltura, Regione Veneto, Venezia, pp. 275-402.

FAO/WHO, 2001, *Codex Alimentarius Commission*, 24th session, Geneva, July; http://www.codexalimentarius.net.

Foreign Agricultural Service (FAS), 2004, online data; http://www.fas.usda.gov.

Gallas, P., 2000, Un mercato in costruzione, *Largo consumo*, 2: 22–34.

Gardner, B., 2000, Organic Food in the European Union: Production, Consumption and the Development of Markets, *Agra Europe*, London; http://www.agra-europe.it.

IFOAM, 2002, *Basic Standards for Organic Production and Processing*, Ifoam General Assembly, Basel, Switzerland; http://www.ifoam.org

International Trade Center (ITC) UNCTAD/WTO, 2001–2004, online data; http://www.intracen.org/home.htm

ISMEA, 2004, Lo scenario economico dell'agricoltura biologica, Roma.

ISMEA, 2002, *La spesa per i prodotti biologici confezionati: 2002*, Osservatorio consumi, Roma; http:www.ismea.it.

Lunati, F., 2002, *Il biologico in cifre 2002*, Rapporti Biobank, Distilleria EcoEditoria, Forlì.

Lunati, F., 2004, *Il biologico in cifre 2004*, Rapporti Biobank, Distilleria EcoEditoria, Forlì.

Pinton, R., 2001, *Some notes about organic market*, paper presented at the "Organic Food and Farming: towards partnership and action in Europe" Conference held in Copenaghen 10–11 May.

Santucci, F.M., and Pignataro, F., 2002, *Organic farming in Italy*, paper presented at the OECD workshop on organic agriculture held in Washington D.C., September 23–26, 2002.

Shaota, A., 2004, Overview of the global market for organic food and drink, in: *The world of organic agriculture*, M. Yussefi and H. Willer, eds., Stiftung Ökologie & Landbau (SÖL), Bad Dürkheim, Germany, pp. 21–26; http://www.soel.de.

USDA Agricultural Marketing Service, 2000, *National Organic Program: Final Rule*, Washington DC, USA; http://www.ams.usda.gov/nop.

USDA-ERS, 2002, Harmony between agriculture and the environment: current issues, http://www.ers.usda.gov.

Yussefi, M., and Willer, H., 2004, *The World of organic agriculture: statistics and emerging trends*, Stiftung Ökologie & Landbau (SÖL), Bad Dürkheim, Germany; http://www.soel.de.

AN OVERVIEW OF ORGANIC AGRICULTURE IN THE UNITED STATES

Catherine Greene[*]

SUMMARY

Organic agriculture has a rapidly growing consumer demand in the United States, and a rapidly developing support infrastructure, making it a premier technology in the efforts of many public and private organizations that advocate more sustainable farming practices. In October 2002, The U.S. Department of Agriculture (USDA) implemented national organic standards on organic production and handling, following more than a decade of development, and over 90 state and private organizations have now been accredited by USDA to certify organic farmers, ranchers, distributors, processors and manufacturers. Although organic grain crop acreage is still under 0.5 percent of the U.S. total, the share is much higher for the fruit, vegetable, and dairy sectors. As the U.S. organic farm sector expands, university-based research and technical assistance and other State and federal support for organic farmers are also beginning to emerge.

1. INTRODUCTION

Farmers have been developing organic farming systems in the United States for over half a century, and organic markets have emerged and expanded greatly during this period. The U.S. Department of Agriculture (USDA) implemented national organic standards on organic production and processing in October 2002, following more than a decade of development, and the new uniform standards are expected to facilitate further growth in the organic farm sector. USDA's organic standards incorporate cultural, biological, and mechanical practices that foster cycling of resources, ecological balance, and protection of biodiversity.

Consumer demand for organically produced goods began to rise sharply during the 1990s, providing market incentives for farmers across a broad range of products. In

---

[*] Agricultural Economist in the Rural and Resource Economics Division of the Economic Research Service, U.S. Department of Agriculture. The views expressed in this article do not necessarily represent those of USDA.

2003, organic retail sales in the United States were estimated at nearly USD 11 billion, approximately 2 percent of total U.S. food sales (Nutrition Business Journal, 2004). The United States market accounted for nearly half of the combined retail sales of organic food and beverages in major world markets in 2003. The UN/WTO International Trade Centre estimated the world organic market at USD 23–25 billion in 2003, slightly more than double estimated world sales in 1997 (ITC, 2002).

An increasing number of farmers in the United States are adopting organic farming systems in order to lower input costs, conserve nonrenewable resources, capture high-value markets, and boost farm income. The U.S. Department of Agriculture (USDA) estimates that farmers and ranchers added a million acres of certified organic land for major crops and pasture between 1997 and 2001, more than doubling organic cropland for major crops (Greene and Kremen, 2003). By 2001, total certified organic cropland and pasture encompassed 2.3 million acres in 48 States. Organic agricultural imports, estimated at between USD 1 billion and USD 1.5 billion in 2002, have also played a significant role in the U.S. market expansion for organic products (USDA, 2005).

The United States ranked fourth in land area managed under organic farming systems, behind Australia (with 24 million acres under organic management), Argentina (7.3 million acres), and Italy (3 million acres) in a 2004 worldwide survey but was not among the top ten countries as a percentage of total farmland under organic management (Yussefi and Willer, 2004). Worldwide conversion levels are currently the highest in European Union countries, which have been providing direct financial support to producers for conversion since the late 1980's to capture the environmental benefits of these systems and support rural development. While government intervention in the United States has focused primarily on market facilitation, several States have begun providing support for conversion to organic farming systems as a way to capture the benefits of these systems (Plank, 1999; DeWitt, 1999).

The rest of this chapter will review in more detail the development of organic farming agriculture in the United States, including the current status of U.S. organic standards and certification requirements and the economic characteristics of the U.S. organic sector, as well as the growing availability of technical assistance in broadening the adoption of organic farming systems.

2. U.S. ORGANIC STANDARDS AND CERTIFICATION

Private organizations, mostly nonprofit, began developing certification standards and third-party certification services in the early 1970's as a way to support organic farming and prevent consumer fraud. Some States began offering organic certification services in the late 1980's for similar reasons. The resulting patchwork of standards in the various certification programs, however, caused a variety of market problems, and the federal government started the process toward uniform standards in 1990.

2.1. U.S. National Organic Standards

Congress passed the Organic Foods Production Act of 1990 to establish national standards for organically produced commodities, and USDA promulgated final rules for implementing this legislation in December 2000, with an 18-month transition period. As

Figure 1. USDA organic seal

of October 2002, all agricultural products that are sold, labeled, or represented as organic in the U.S. are to be in compliance with the regulations, and the USDA organic seal[1] may be applied to certain products (Figure 1). The regulations require that organic growers and handlers (including food processors and distributors) be certified by a State or private group under the uniform standards developed by USDA, unless the farmers and handlers sell less than USD 5,000 a year in organic agricultural products. Retail food establishments that sell organically produced agricultural products, but do not process them, are also exempt from certification.

Organic farming systems rely on ecologically based practices such as biological pest management and composting; virtually exclude the use of synthetic chemicals, antibiotics, and hormones in crop production; and prohibit the use of antibiotics and hormones in livestock production. Under organic farming systems, the fundamental components and natural processes of ecosystems—such as soil organism activities, nutrient cycling, and species distribution and competition—are used as farm management tools. In USDA's final national organic rule, organic production is defined as "a production system that is managed in accordance with the Act and regulations in this part to respond to site-specific conditions by integrating cultural, biological, and mechanical practices that foster cycling of resources, promote ecological balance, and conserve biodiversity" (U.S. Department of Agriculture, 2002).

While the National Organic Food Production Act of 1990 did not target improvements in environmental and human health as an explicit objective of the regulation, these concerns are addressed in Section 2119 of the Act which establishes the following set of

---

[1] USDA rules include provisions for enforcement. USDA organic labeling requirements apply to raw, fresh, and processed products that contain organic ingredients and are based on the percentage of organic ingredients in a product.

Agricultural products labeled "100 percent organic" must contain (excluding water and salt) only organically produced ingredients. Products labeled "organic" must consist of at least 95-percent organically produced ingredients. Products labeled "made with organic ingredients" must contain at least 70-percent organic ingredients. Products with less than 70-percent organic ingredients cannot use the term organic anywhere on the principal display panel but may identify the specific ingredients that are organically produced on the ingredients statement on the information panel.

The USDA organic seal—the words "USDA organic" inside a circle—may be used on agricultural products that are "100 percent organic" or "organic". A civil penalty of up to USD 10,000 per violation can be levied on any person who knowingly sells or labels as organic a product that is not produced and handled in accordance with the regulations.

criteria for approving and prohibiting substances for use in organic production and handling operations: (1) the potential of such substances for detrimental chemical interactions with other materials used in organic farming systems; (2) the toxicity and mode of action of the substance and of its breakdown products or any contaminants, and their persistence and areas of concentration in the environment; (3) the probability of environmental contamination during manufacture, use, misuse or disposal of such substance; (4) the effect of the substance on human health; (5) the effects of the substance on biological and chemical interactions in the agroecosystem, including the physiological effects of the substance on soil organisms (including the salt index and solubility of the soil), crops and livestock; (6) the alternatives to using the substance in terms of practices or other available materials; and (7) its compatibility with a system of sustainable agriculture.

The national organic standards address the methods, practices, and substances used in producing and handling crops, livestock, and processed agricultural products. Although specific practices and materials used by organic operations may vary, the standards require every aspect of organic production and handling to comply with the provisions of the Organic Foods Production Act of 1990. Organically produced food cannot be produced using genetic engineering and other excluded methods, sewage sludge, or irradiation. These standards include a national list of approved synthetic substances such as insecticidal soaps and horticultural oils, and prohibited non-synthetic substances (such as arsenic, strychnine, and tobacco dust) for use in organic production and handling.

Organic livestock production systems attempt to accommodate an animal's natural nutritional and behavioral requirements, ensuring that dairy cows and other ruminants, for example, have access to pasture. USDA organic livestock standards incorporate requirements for living conditions, pasture and access to the outdoors, feed ration, and health care practices suitable to the needs of the particular species.

2.2. U.S. Organic Accreditation, Certification and Eco-labeling

The number of domestic organizations offering certification services to U.S. growers has grown steadily over the last decade, both during and after implementation of national organic rules in 2002 requiring certifiers to be accredited to meet national organic standards. By 2005, the U.S. Department of Agriculture had accredited fifty-six State and private organizations in the United States to provide certification services to organic farmers, and had also accredited more that forty certification organizations in other countries to certify organic products to U.S. standards.

Many of the oldest private certifiers in the U.S.—including Northeast Organic Farming Association of Vermont, Maine Organic Farmers and Gardeners Association, and California Certified Organic Farmers—obtained USDA accreditation along with newer groups. The private organizations that offer certification services in the U.S. are still mostly nonprofit organizations.

USDA's national standards do not restrict additional eco-labeling of organic products, and some organic certifiers are also developing standards on social and other aspects of agricultural production and food distribution—such fair trade and local sourcing. The Florida Certified Organic Growers and Consumers organization, for example, recently developed a partnership program for food retailers and restaurants in North Florida to certify their level of commitment to local food sourcing. Most coffee sold in the

U.S. that is certified to guarantee farmers get a fair price also has a separate organic certification, and some certification groups are trying to improve the efficiency of their certification efforts by integrating these programs.

A number of U.S. organic certifiers also perform certification services for U.S. growers who want to meet international standards, such as International Federation of Organic Agriculture Movements Norms. In 2003, the U.S. National Institute of Standards and Technology (NIST) established a component for organic production and processing under its National Voluntary Conformity Assessment System Evaluation program. In 2004, NIST issued independent recognition to a U.S. based nonprofit, the International Organic Accreditation Service, to accredit organic certification bodies for compliance with several international standards, including IFOAM norms.

3. ECONOMIC CHARACTERISTICS OF THE U.S. ORGANIC AGRICULTURE SECTOR

Organic farming has a long history in the U.S., and its recent rapid growth may have increased risks for some organic farmers. Between 1997 and 2001, U.S. farmers and ranchers added about one million acres of certified organic cropland and pasture, and the number of farming operations with certified organic acreage, excluding subcontractors, increased from about 5,000 to nearly 7,000 (Greene and Kremen, 2003). The average size of certified organic farm operations increased as well during this period, as existing organic farmers expanded their operations and new large-scale operations became certified. Before this recent rapid growth, price premiums associated with "organic niche markets and 'family farms' were said to be at risk when large-scale organic producers or processors enter the market, if demand does not expand sufficiently" (Dobbs *et al.*, 2000). However, although organic price data is not widely available for most commodities, several recent studies on produce and grains have not show substantial declines in price premiums since the mid-1990s (Streff and Dobbs, 2004; Oberholtzer *et al.*, 2005).

3.1. Organic farm size and marketing arrangements

J.I. Rodale began popularizing organic farming systems in the U.S. in the 1940s with the publication of *Organic Farming and Gardening* magazine, and a few farmers began experimenting with these systems, marketing directly to consumers (Kelly, 1992). By the late 1950s organic foods were being featured in small health food stores. By the late 1960s, "a new generation of environmentally conscious consumers—Baby Boomers—were coming of age and demanding foods produced without chemicals" (Mergentime, 1994). Large natural foods supermarkets featuring organic foods began developing in the 1980s, and by the end of the 1990s, organic products were becoming widely available in conventional supermarkets as well.

Organic products are now available throughout the mainstream food chain, including nearly 20,000 natural food stores and 73 percent of conventional grocery stores, according to recent industry statistics (Dimitri and Greene, 2002). The expansion of organic products into mainstream marketing venues has opened opportunities for some farmers interested in producing for larger markets. Farmers' markets and other direct-market venues have also grown in number over the last decade, and continue to be especially popular among organic producers, (Kremen *et al.*, 2004). U.S. organic farmers are also finding ways to capture a larger segment of the consumer food dollar through on-farm processing, producer marketing cooperatives, and new forms of direct marketing, including "community supported agriculture" (CSA) farm subscription services.

In the United States, California had the most certified organic cropland in the U.S. in 2001, and has the nation's largest concentration of fruit and vegetable producers, both conventional and organic. Northeastern States have a relatively small amount of organic cropland, but have a large concentration of organic market gardeners. Most of the organic acreage in the North central and upper Midwestern States is used for grain, soybean, and oilseed production. Certified organic pasture and ranchland was concentrated in three western States, although over 40 States had some certified organic pasture in 2001. Most states in the Southeastern U.S. had very little certified organic cropland, pasture or operations.

Average size of certified organic farm operations is up for the U.S. as a whole, as existing organic farmers expand their operations and new large-scale operations become certified, but small-scale farms remain the prevalent organic operation. In California, where the majority of large organic fruit and vegetable operations are located, the size of certified organic operations began increasing in the 1980's and the average size more than tripled between 1985 and 1991 (Greene, 1992). Still, a decade later most of the organic farms remain small. Recent farm trends by the University of California indicates that the state's organic farms remained small (under 5 acres on average) throughout the late 1990s (Klonsky *et al.*, 2002).

3.2. Risks in organic farming

Organic farming, which is distinguished from conventional farming by its reliance on the natural processes of ecosystems, may present particular risks and ways of managing risks. Organic farming systems virtually exclude what are often thought of as important risk management tools in conventional farming, such as the use of synthetic chemicals and antibiotics. Instead, organic farmers rely on their understanding and management of cultural practices such as crop rotation, timing of planting and harvesting, mechanical cultivation, and development of beneficial insect populations.

Several studies of the characteristics of organic and sustainable farms have included an examination of risk management issues. Organic farmers expose themselves to special risks during the 3-year transition period from conventional to organic (Duram, 1999). Yields may drop, and transitioning producers are not able to obtain the higher prices that come with certified organic production. Also during the transition, farmers are learning a different way of farming, potentially with a different set of crops, building up the organic matter in their soil, and identifying new sources of information and inputs. Sustainable agriculture farmers in Kansas referred to the transition period as a time of "thinking risks" (Hanson *et al.*, 1990).

Concerns about personal risk from occupational pesticide exposure and environmental contamination may be especially important to organic farmers. The Environmental Protection Agency estimates that 10,000–20,000 physician diagnosed pesticide poisonings occur each year among approximately 3,380,000 U.S. agricultural workers (National Institute for Occupational Safety and Health, 2004). The extent of chronic illness resulting from pesticide exposure is much less documented. Epidemiological studies of cancer suggest that farmers in many countries, including the United States, have higher rates than the general population for Hodgkin's disease, leukemia, multiple myeloma, non-Hodgkin's lymphoma, and cancers of the lip, stomach, prostate, skin, brain, and connective tissue (Blair and Zahm, 1991). Some case reports and experimental studies suggest that pesticide exposure is a risk factor for neurodegenerative diseases on other non-cancer illnesses (Alavanja *et al.*, 1993). Many of the organic farmers profiled in alternative agricultural studies include concerns about chemical use as motivations for farming organically (Thrupp, 2002; Sustainable Agriculture Research and Education Program, 2001).

The Organic Farming Research Foundation (OFRF), a California-based nonprofit, has conducted several national surveys of organic producers since 1993, and has identified a number of production risks in organic farming, including serious shortages of some organic inputs at times (OFRF). A research team from the University of Maryland and the USDA Economic Research Service conducted focus groups with organic farmers from different regions in the United States in 2001 and 2002, and found that some risks in organic farming, such as weather and climatic risks, are similar to those in conventional farming, although they might be managed in different ways. Other risks, such as growing pains with the implementation of national organic standards, are related specifically to organic regulation and marketing. Organic farmers are especially vulnerable to variable prices because the organic market is immature, and to accidental contamination of their land because they make relatively large investments in soil quality (Hanson *et al.*, 2004).

Growth in the organic sector has spurred government interest in the risk management needs of organic producers. Legislation passed by the U.S. Congress, the Agricultural Risk Protection Act of 2000, recognized organic farming as a "good farming practice" that would be covered by Federal crop insurance. Federal crop insurance began covering transitional and certified organic acreage the following year under written agreements.

3.3. Productivity and profitability in organic farming systems

A limited, but growing, number of studies in the United States have examined the yields, input costs, profitability, managerial requirements, and other economic characteristics of organic farming. A 1990 review of the U.S. literature at Cornell concluded that the "variation within organic and conventional farming systems is likely as large as the differences between the two systems," and found mixed results in the comparisons for most characteristics (Knoblauch *et al.*, 1990).

Several USDA and university studies during the 1990's in California, Ohio, and Texas have indicated that organic price premiums are necessary to give organic farming systems comparable or higher whole-farm profits than conventional chemical-intensive systems, particularly for crops like processed tomatoes and cotton. A Wallace Institute of Alternative Agriculture review of university-based comparative studies in the 1980's and early 1990's on Midwestern organic grain and soybean production found organic systems needed price premiums to be more profitable than conventional systems (Welsh, 1999). Several of these studies, however, found that organic grain and soybean production could be as profitable even without price premiums due to higher yields in drier areas or periods, lower input costs, or higher revenue from the mix of crops used in the system. Other recent studies have also found that some organic systems may be more profitable than

Table 1. Examples of U.S. multidisciplinary, long-term research projects with organic trials

Project	Date Established	Farming system/Commodity focus
University of Nebraska-Lincoln Long-Term Experiment Trials	1975	Compare conventional and organic systems (Rotations include corn, wheat, and soybean)
Rodale Institute—Kutztown, PA Farming Systems Trial™	1981	Examine the transition process from conventional to organic farming (corn and soybeans)
University of California-Davis Sustainable Agriculture Farming Systems Project	1988	Compare conventional, low-input and organic systems; evaluate conservation tillage in these systems (tomato, safflower, bean, corn)
Iowa State Univ.—Leopold Center Neely-Kinyon Long-Term Agroecological Research	1988	Compare conventional and organic systems (corn, soybeans and alfalfa)
University of Minnesota-Southwest Research and Outreach Center Elwell Agroecology Farm	1989	Compare conventional and organic systems (corn, soybeans, alfalfa and oats)
Michigan Agricultural Experiment Station Living Field Laboratory	1993	Compare conventional, organic and other systems (corn, soybeans, and wheat)
USDA Agricultural Research Center-Beltsville, MD Farming Systems Project	1993	Compare organic systems typical in the mid-Atlantic region (corn and soybeans)
West Virginia University (WVU) Horticulture Farm Project	1999	Evaluate organic systems on the entire Horticulture Farm (market garden and field crop/livestock systems)
North Carolina State University Farming Systems Trial	2001	Compare conventional and transitional organic systems (grains, livestock and woodlots)
Ohio State University John Hirzel Sustainable Agriculture Research and Education Site	2001	Compare conventional, no-till, and high- and low-input organic systems (soybeans, corn, wheat, and vegetables)

Source: USDA Economic Research Service.

conventional systems, even without price premiums (Swezey et al., 1994; Reganold et al., 2001).

Net returns to both conventional and organic production systems vary with biophysical and economic factors such as soil type, climate, proximity to markets, and other factors that are farm specific, and help explain the wide variation in economic performance within each system. Factors not captured in standard profit calculations—such as convenience, longer-term planning horizons, and environmental ethics—can motivate rational adoption of a particular practice or farming system. Our understanding of the factors influencing net returns to organic farming systems remains imperfect.

Table 2. Adoption rates for certified organic farming systems in the U.S., by crop (2001)

Specialty Crop	Share of U.S. acreage	Field	Share of U.S. acreage
Tomato	1.0%	Corn	0.1%
Grape	1.5%	Soybean	0.2%
Apple	3.0%	Wheat	0.3%
Carrot	4.0%	Oat	0.8%
Lettuce	5.0%	Rice	1.0%

Source: USDA Economic Research Service.

Economic research on organic farming has tended to focus narrowly on profitability (Fox et al., 1991), but land-grant universities and others are increasingly examining the long-term economics of organic systems through replicated field trial research and a multidisciplinary systems approach (Table 1). According to the Organic Farming Research Foundation, 18 states had land grant institutions with research acres under certified organic management in 2003, up from only six states in 2001 (Sooby, 2003). Organic farming systems trials—in experiment stations and on-farm settings—promise to answer basic research questions about yields, profitability, and environmental impacts, as well as to address farmer-defined management and production obstacles to adoption of organic production systems.

3.4. Adoption of organic farming systems in the U.S.

American farmland under organic management has grown steadily for the last decade as farmers strive to meet consumer demand in both local and national markets. U.S. certified organic crop acreage more than doubled between 1992 and 1997, and has doubled again between 1997 and 2001 for most major crops (Greene and Kremen, 2003). Certified organic pasture and ranchland also doubled between 1997 and 2001, following USDA's lifting of restrictions on organic meat labeling in the late 1990's.

U.S. farmers and ranchers have added another million acres of certified organic cropland and pasture since 1997, bringing the 48-State total to 2.34 million acres in 2001. Certified organic livestock grew even faster during this period. Most crop/livestock sectors and most States also showed strong growth between 2000 and 2001. Overall, certified organic cropland and pasture accounted for 0.3 percent of U.S. cropland and pasture in 2001, although the share is much higher in some crops, such as vegetables at over 2 percent (Table 2).

California was the leading State in certified organic acreage in 2001, with nearly 150,000 acres, mostly used for fruit and vegetable production. North Dakota followed closely with nearly 145,000 acres, mostly for wheat, soybeans, and other field crops. Minnesota, Wisconsin, Iowa, and Montana were other top States.

Over 40 States had certified pasture and rangeland in 2001, most with under 20,000 acres, although several States had over 100,000 acres. The number of certified organic beef cows, milk cows, hogs, pigs, sheep, and lambs was up nearly four-fold since 1997, and up 27 percent between 2000 and 2001. Dairy has been one of the fastest growing segments of the organic foods industry during this period, and milk cows accounted for over half of the certified livestock animals. Poultry animals raised under certified organic management—including layer hens, broilers, and turkeys—showed even higher levels of growth during this period.

4. RECENT STATE AND FEDERAL POLICY INITIATIVES

Government research and policy initiatives often play a key role in the adoption of new farming technologies and systems. Worldwide adoption levels for organic farming systems are currently the highest in European Union countries, where governments have been developing consumer education initiatives and providing direct financial support to producers for conversion since the late 1980s to capture the environmental benefits of these systems and support rural development.

Federal support for organic farmers and handlers is also beginning to emerge in the United States. USDA agencies have started or expanded programs on organic agriculture during the last several years, and the Farm Security and Rural Investment Act of 2002 (the 2002 Farm Act) contains several first-time research and technical assistance provisions to directly assist organic crop and livestock producers with production and marketing (Greene and Kremen, 2003). These provisions reflect the potential benefits from organic farming systems which include improved soil tilth and productivity, lower energy use, and reduced use of pesticides. Recent programs and initiatives include:

- *Certification Cost-Share Support.* In 2001, USDA established a certification cost-share program to help farmers defray certification costs in 15 States, and the 2002 Farm Act allocated USD 5 million in cost-share assistance funds for this program and expanded eligibility to growers and handlers in all States.
- *Research and Technical Assistance.* The 2002 Farm Act contains an Organic Agriculture Research and Extension Initiative that authorizes USD 3 million per year in new mandatory appropriations in fiscal years 2003–2007. These funds are being used to administer competitive research grants focused on organic agriculture production, breeding, and processing methods and the marketing and policy constraints in organic agricultural sector.
- *Conservation Initiatives.* The 2002 Farm Act provided funding for the Conservation Security Program, which provides payments to producers for adopting or maintaining land management and conservation practices to address resource concerns. This new program may interest organic farmers who commonly adopt these types of practices as part of their organic farming systems.
- *Marketing Order Exemptions.* Another provision in the 2002 Farm Act specifies that certified organic producers who produce and market only organic products and do not produce any conventional or nonorganic products will now be exempt from paying an assessment under any commodity promotion law. USDA published proposed rule to implement this legislation in December 2003.
- *Export Promotion, Crop Insurance, and Other Initiatives.* USDA's Risk Management Agency has provided insurance coverage for organic farming practices as good farming practices by written agreement since 2001, and is working to improve its organic crop insurance program. USDA's Foreign Agricultural Service (FAS) is helping design protocols for working with foreign nations to keep organic trade moving as more countries develop organic standards.

State support for organic farmers and handlers is also beginning to emerge in the United States. Minnesota and Iowa, for example, began subsidizing conversion to organic farming systems in the late 1990s as a way to capture the environmental benefits of these systems, and several other states have begun similar programs since then. In 2003, the National Association of State Departments of Agriculture (NASDA) released a policy statement on organic agriculture expressing support for a wide range of activities that would expand public-sector organic research and education and provide technical assistance to organic and transitional farmers (NASDA).

5. REFERENCES

Alavanja, M.C.R., Blair, A., McMaster, S.B., and Sandler, D.P., 1993, *Agricultural health study: a prospective study of cancer and other diseases among mean and women in agriculture*. National Cancer Institute, U.S. Environmental Protection Agency, and the National Institute of Environmental Health Sciences, Oct. 25 (Revisions Dec. 16).

Blair, A., and Zahm, S.H., 1991, Cancer among farmers, *Occupational Medicine: State of the Art Reviews*, **6**(3): 335–354.

DeWitt, J., 1999, Iowa State University Extension, personal communication. [Also, see Jerry DeWitt, "Organic Production and EQIP," May 29, 1997, in USDA Sustainable Agriculture Network (SANET-mg) archives; www.sare.org/htdocs/hypermail/index.html.]

Dimitri, C., and Greene, C., 2002, *Recent Growth Patterns in the U.S. Organic Foods Market*, Agriculture Information Bulletin No. 777, U.S. Department of Agriculture, Economic Research Service, September.

Dobbs, T.L., Shane, R.C., and Feuz, D.M., 2000, Lessons learned from the Upper Midwest organic marketing project, *American Journal of Alterative Agriculture*, **15**(3): 119–128.

Duram, L.A., 1999, Factors in organic farmers' decision making: Diversity, challenge, and obstacles, *American Journal of Alternative Agriculture*, **14**(1): 2–10.

FOG/QCS, 2003, *Certification Manual of North Florida Local Food Partnership, 2003–2004*. Florida Organic Growers & Consumers/Quality Certification Services, Gainesville, FL.

Fox, G., Weersink, A., Sarwar, G., Duff, S., and Deen, B., 1991, Comparative Economics of Alternative Agricultural Production Systems: A Review, *Northeastern Journal of Agricultural and Resource Economics*, **20**(1): 124–142.

Greene, C., and Kremen, A., 2003, *U.S. Organic Farming in 2000–2001: Adoption of Certified Systems*. Agriculture Information Bulletin 780. Economic Research Service, U.S. Department of Agriculture. February.

Greene, C., 1992, *Steady Success in Organic Produce*, Agricultural Outlook, AO-185, Economic Research Service, U.S. Department of Agriculture, May.

Hanson, J.C., Johnson, D.M., Peters, S.E. and Janke, R.R., 1990, The profitability of sustainable agriculture on a representative grain farm in the Mid-Atlantic Region, 1981–1989. *Northeastern Journal of Agricultural and Resource Economics*, **19**(2): 90–98.

Hanson, J., Dismukes, R., Chambers, W., Greene, C., and Kremen, A., 2004, Risk and risk management in organic agriculture: Views of organic farmers, *Renewable Agriculture and Food Systems*, **19**(4): 218–227.

International Trade Centre UNCTAD/WTO (ITC), 2002, Overview: World Markets for Organic Food and Beverages, ITC, Geneva; www.intracen.org/mds/sectors/organic/.

Kelly, W.C., 1992, Rodale Press and Organic Gardening, in Proceedings of the Workshop on the History of the Organic Movement, American Society for Horticultural Science, *HortTechnology*: **2**(2), 270–271.

Klonsky, K., Tourte, L., Kozloff, R., and Shouse, B., 2002, *A Statistical Picture of California's Organic Agriculture: 1995–1998*, University of California Agricultural Issues Center, August.

Knoblauch, W.A., Brown, R., and Braster, M., 1990, *Organic Field Crop Production: A review of the Economic Literature*, A.E. Res. 90-10, Department of Agricultural Economics, Cornell University, Ithaca, NY, July.

Kremen, A., Greene, C., and Hanson, J., 2004, *Organic Produce, Price Premiums, and Eco-Labeling in U.S. Farmers' Markets*, Outlook Report No. (VGS-301-01), USDA, Economic Research Service, 12 pp, April.

Mergentime, K., 1994, Organic Industry Roots Run Deep, *Natural Foods Merchandiser*, June.

National Institute for Occupational Safety and Health., 2006, NIOSH Safety and Health Topic: Pesticide Illness & Injury Surveillance. Centers for Disease Control and Prevention, Department of Health and Human Services; www.cdc.gov/niosh/topics/pesticides/.

Nutrition Business Journal, 2004, *NBJ's Organic Foods Report 2004*, Penton Media/New Hope Natural Media, San Diego, CA.

Oberholtzer, L., Dimitri, C., and Greene, C., 2005, *Organic Price Premiums Hold on as Organic Produce Market Expands*, Outlook Report No. VGS-30801, Economic Research Service, U.S. Department of Agriculture, May.

Organic Farming Research Foundation, 1999, *Final results of the third biennial (1997) national organic farmers' survey*. Organic Farming Research Foundation, Santa Cruz, CA.

Plank, D., 1999, Minnesota Passes Law to Help Organics, *Natural Foods Merchandiser*, August.

Reganold, J.P., Glover, J.D., Andrews, P.K., and Hinman, H.R., 2001, Sustainability of three apple production systems, *Nature*, 410: 926–930.

Sooby, J., 2003, *State of the States 2nd Edition: Organic Farming Systems Research at Land Grant Institutions 2001–2003*. Organic Farming Research Foundation, Santa Cruz, CA.

Streff, N., and Dobbs, T.L., 2004, *"Organic" and "Conventional" Grain and Soybean Prices in the Northern Great Plains and Upper Midwest; 1995 through 2003*, South Dakota State University Economics Pamphlet 2004-1, Brookings, SD, June.

Sustainable Agriculture Research and Education (SARE) program, 2001, *The new American farmer: profiles of agricultural innovation*, Cooperative State Research, Education and Extension Service, USDA, pp. 159; www.sare.org/naf/index.htm.

Swezey, S., Rider, J., Werner, M., Buchanan, M., Allison, J., and Gliessman, S., 1994, Granny Smith conversions to organic show early success in Santa Cruz County, *California Agriculture*, **48**(6): 36–44.

Thrupp, L.A., 2002, *Fruits of progress: growing sustainable farming and food systems*, World Resources Institute, Washington, DC.

U.S. Department of Agriculture, 2000, *National Organic Program, final Rule*, Federal Register 7 CFR Part 205, December 21 [see www.ams.usda.gov/nop].

U.S. Department of Agriculture, 2002, *Veneman Marks Implementation of USDA National Organic Standards*, USDA News Release No. 0453.02, October 21.

U.S. Department of Agriculture, Foreign Agriculture Service, 2005, *U.S. Market Profile for Organic Food Products*. February.

Yussefi, M., and Willer, H., 2004, *The World of organic agriculture: statistics and emerging trends*, 6th revised edition, Stiftung Ökologie & Landbau (SÖL), Bad Dürkheim, Germany; http://www.soel.de.

Welsh, R., 1999, *The Economics of Organic Grain and Soybean Production in the Midwestern United States*, Policy Studies Report No. 13, Henry A. Wallace Institute for Alternative Agriculture, May.

THE PRODUCER'S PERSPECTIVE

A COMPARATIVE PROFITABILITY ANALYSIS OF ORGANIC AND CONVENTIONAL FARMS IN EMILIA-ROMAGNA AND IN MINNESOTA

Maurizio Canavari, Rino Ghelfi, Kent D. Olson, and Sergio Rivaroli[*]

SUMMARY

Recent discussion surrounding organic agriculture (also referred to as organic farming) has turned from just whether it represents a viable alternative to conventional agriculture to whether it would be adopted by a significant percentage of farmers. After a beginning phase in which the adoption was mainly due to an ethically based choice of the farmer, the success in the market and the increasing demand for organic products are increasing the number of farmers converting their farming system. Despite the continuing importance of non-economic factors and the uncertainty given by short-term and mid-term fluctuations of prices, a decisive point is whether the conversion to organic farming may be worthwhile from an economic perspective. The aim of the paper is to compare the actual profitability of farms using organic production methods to those farms using conventional production methods. The analysis will be based on several data-sets, provided on the Italian side by Emilia-Romagna Region, Italian National Institute of Agricultural Economics (INEA), Italian National Institute of Statistics (Istat), and on the US side by the Center for Farm Financial Management (CFFM) in the Department of Applied Economics, University of Minnesota.

1. INTRODUCTION

The advent (or return) of organic agriculture has been listed as one of the many innovations affecting and transforming the food industry today (Santucci, 2002a; Santucci and Chiorri, 1996). To its proponents, organic agriculture represents an alternative to conventional cultivation system.

---

[*] Maurizio Canavari, Rino Ghelfi, and Sergio Rivaroli, Dipartimento di Economia e Ingegneria Agrarie, Alma Mater Studiorum-Università di Bologna. Kent Olson, Department of Applied Economics, University of Minnesota. Rino Ghelfi: sections 1 and 2. Maurizio Canavari: section 3. Sergio Rivaroli: sections 4.1 and 4.2. Kent Olson: sections 4.3 and 5.

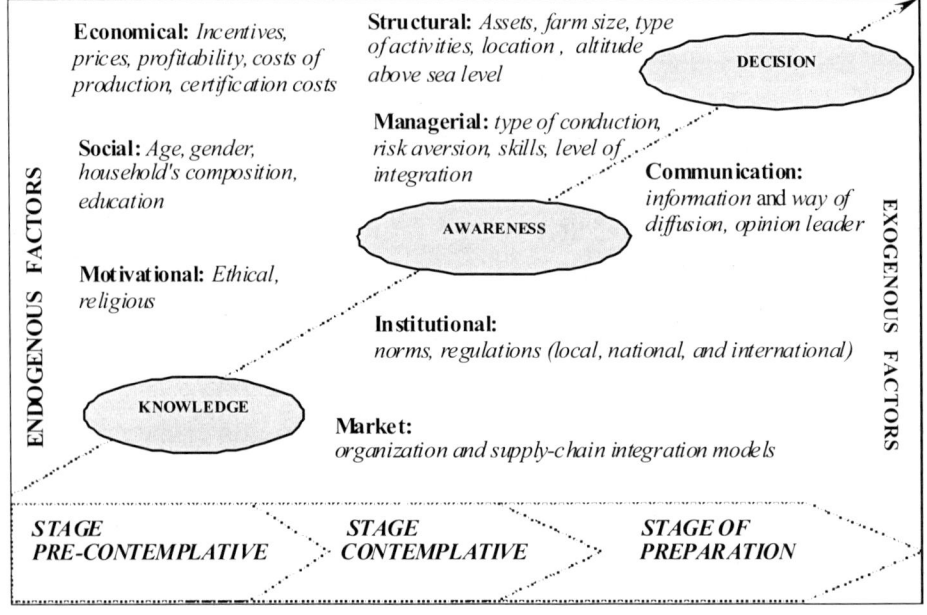

Figure 1. The innovation-adoption decision process

The adoption of an innovation (such as organic farming) may be interpreted as a mental process leading the agricultural entrepreneur to form an opinion on the opportunity to make or not to make a certain choice (Rogers, 1983). According to Prochaska et al. (1992), this mental evolution develops through phases characterizing the decision path of the entrepreneur. Figure 1 shows a typical decision path leading to acceptance or refusal of innovation and the influences of endogenous and exogenous factors. The endogenous factors are linked to changes in structural, managerial, socio-economic characteristics of the entrepreneur and of his or her family. The exogenous factors are linked to changes in institutions, infrastructures, socio-economic characteristics of the business environment and changes in demand and supply conditions within the market.

After a beginning phase in which the adoption of organic farming was mainly due to an ethically based choice of the farmer, the number of farmers converting to an organic farming system is increasing due to the early success in the market and the increasing demand for organic products. However, due to the continuing presence and importance of non-economic factors and the uncertainty given by short-term and mid-term fluctuations of prices, farmers are still debating whether the conversion to organic farming may be worthwhile from an economic perspective. Thus, the question remains as to how many farmers would convert to organic farming systems.

After the analysis of the influence of structural factors into the farmer's decision process, the aim of the paper is to compare the actual and potential profitability of farms using organic production methods to those farms using conventional production methods.

2. OBJECTIVES AND HYPOTHESES

The conceptual framework just described highlights the fact that the conversion to organic agriculture may be influenced by a set of various factors. While it would be more desirable to jointly consider many elements that may influence a farmer's choice, the information available only allows us to test the influence of structural factors on the decision path followed by a group of agricultural enterprises in the Emilia-Romagna region. Based on the conceptual framework and the available data, the following hypotheses are tested.

H_1: *structural characteristics of the farm are related to the adoption of organic farming;*

H_2: *a direct relationship exists between income level and organic farming adoption.*

A third objective is to compare the profitability of organic and conventional farms in Emilia-Romagna and in Minnesota.

3. MATERIALS AND METHODS

The first part of this analysis was carried out using administrative and accounting data for 1,781 balance sheets of Emilia-Romagna farms from 2000 through 2003. This database consists of the bookkeeping data from a sample of farms within of the European Union Network of Farm Accountancy Data (FADN). The information can be considered as representative of an average situation for the 2000–2003 period.

For the profitability comparison between organic and conventional farms, a subgroup was selected from the whole sample consisting of those farms using organic farming methods and receiving EU subsidies for sustaining this practice. While homogeneous in terms of organic farming practices, these farms are different in terms of size, type of farmers, and so on. The accountancy data have been standardized, so the economic parameters are consistent and comparable across the sample and also with the group of US farms being used for comparison.

In Minnesota, the data for this report was obtained from the University of Minnesota's Center for Farm Financial Management's (CFFM's) FINBIN© database system. However, since we did not have access to individual observations, comparable farms were obtained by selecting and deselecting size and other characteristics of the organic farms (within FINBIN's reporting system) to obtain as homogeneous a group of farms as possible given the limits in the absolute number of farms classified as organic and the reporting mechanisms within FINBIN. A balance had to be made between having a sufficient number of reporting farms and the breadth of the characteristics. In the four years from 2000 through 2003, there were 42 observations of organic farms: 10 from 2000, 11 from 2001, 10 from 2002, and 11 from 2003. These 42 balance sheets were specified as being predominately crop; dairy; crop and dairy; crop and hog; crop and beef; and other more diversified farms. The conventional farm observations were 2,897; 397 from 2000, 439 from 2001, 996 from 2002, and 1,065 from 2003 for the same types of farms and geographical areas.

The analysis used to test the H_1 and H_2 hypotheses and the profitability comparison is composed of five steps. Each step uses a specific multivariate analysis technique

(Table 1). The first two steps are used to obtain homogenous sets of farms for the comparisons and tests done in the later steps. In the first step, a factor analysis is performed using principal components analysis and the Varimax rotation to extract a smaller number of factors from several structural variables available on the sample farms. This method is chosen to maximize the variability considered, while minimizing the number of variables needed in subsequent analysis. For the second step, the results of the factor analysis in the first step are used to identify homogeneous groups within the Emilia-Romagna sample using a two-step clustering technique that automatically manages categorical variables and selects the optimal number of clusters.

In the third step, the factors affecting the choice to adopt organic farming are analyzed using a logistic regression or LOGIT procedure following other studies of innovation adoption (Jarvis, 1981; Rogers, 1983; Young and Shumway, 1991; Sterns and Peterson, 1996; Neupane *et al.*, 2002; Menard, 2002). Since the Emilia-Romagna data set lacks information on farmer's characteristics, the resulting choice model includes just structural variables:

$$\text{logit}(Y) = \ln\left\{\frac{P_i}{1-P_i}\right\} = \beta_0 + \beta_1 X_1 + \ldots + \beta_i X_i + \varepsilon_i \tag{1}$$

where Y is a (0,1) value, P is the probability that Y assumes a value equal to 1 given the level of the independent variables (in this case the structural factors $X_1, X_2, \ldots, X_i$), β_0 is the intercept and $\beta_1, \ldots, \beta_i$ are the coefficients associated with the independent variables, and εi is the error term. Subscripts indicating individual observations are not specified to improve readability. In this case, the error term needs to be interpreted as the effect of variables not considered in the equation, as well as measurement errors. The H_1 hypothesis is verified if any single independent variable in the logistic regression model and the pseudo R-square index are significant.

In the fourth step, multiple linear regression is used to test the H_2 hypothesis of a relationship among organic farming and profitability:

$$Y_i = \beta_0 + \beta_1 X_1 + \ldots + \beta_i X_i + \varepsilon_i \tag{2}$$

where Y is the income level and $X_1, \ldots, X_i$ are explanatory variables. The hypothesis is verified if the dummy variables describing the adoption of organic farming have significant coefficients. The relationship will be positive or negative according to the sign of the coefficient.

Table 1. Variables and techniques used

Variable	Description	Type	Unit
\multicolumn{4}{c}{Factor analysis (Principal component analysis)}			
X_1 Sau	Farm utilized surface area	Cardinal	Hectare = ha
X_2 Sauirb	Irrigable surface area	Cardinal	ha
X_3 Ucv	Mechanical Power available	Cardinal	Horse Power
X_4 Uuluf	Household's working units	Cardinal	Standard Labor Units
X_5 Uulus	Hired working units	Cardinal	Standard Labor Units
X_6 Macchi	Cost for farm's equipments	Cardinal	EUR/ha
X_7 Nolegg	Cost for external services	Cardinal	EUR/ha
X_8 Debiti	Debts	Cardinal	EUR/ha
X_9 Capfop	Land capital	Cardinal	EUR/ha
X_{10} Capese	Working capital	Cardinal	EUR/ha
X_{11} ALT	Altitude above sea level	Ordinal	meters
Logistic regression			
Bio	Adoption of org. agriculture	Nominal	Yes/No
Factors F1–F4	Structural factors	Cardinal	
Cluster analysis			
Factors F1–F4	Structural factors	Cardinal	
Multiple linear regression			
Factors F1–F4	Structural factors	Cardinal	
Reddit	Return (per hectare)	Cardinal	EUR/ha
BIO_2000	Organic in 2000	Nominal	Yes/No
BIO_2001	Organic in 2001	Nominal	Yes/No
BIO_2002	Organic in 2002	Nominal	Yes/No
BIO_2003	Organic in 2003	Nominal	Yes/No
Rictot	Total sales and subsidies	Cardinal	EUR/ha
SEMINAT	Activity: Arable crops	Nominal	Yes/No
ORTOFRUTT	Activity: Fruit and veg. crops	Nominal	Yes/No
VITE	Activity: Vineyard	Nominal	Yes/No
BOVINI	Activity: cattle	Nominal	Yes/No
SUINI	Activity: hogs	Nominal	Yes/No
P_COLT	Activity: diversified	Nominal	Yes/No
COLT_BES	Activity: mixed	Nominal	Yes/No

Source: Authors' elaboration.

In the fifth step, the economic results of organic and conventional farms in Emilia-Romagna and then in Minnesota are compared using simple statistics. The statistical significance of the differences between the means in the Emilia-Romagna data is tested using a simple F-test.

4. RESULTS AND DISCUSSION

4.1. The choice to adopt organic farming in Emilia-Romagna

In the first step of the analysis, the factor analysis applied to the eleven structural variables listed in Table 1 captured about 55% of their variability using just 4 factors, all having a 99.9% significance level. Following Kaiser (1974), the overall fit of the factor model is confirmed by the value of the KMO test (0.61). This solution is able to explain over the 50% of variance for all the variables except the X_9 (.340) and X_6 (.474) as depicted through the values of communality in Table 2.

In the VARIMAX rotated factor solution, the variables X_8, X_5, X_{10} load significantly on Factor 1 (F1) named "Asset intensity"; the variables X_4, X_3, X_1, X_6 load significantly on Factor 2 (F2) named "Labor intensity"; the variables X_{11}, X_7 load significantly on Factor 3 (F3) named "De-structuration", and the variables X_2, X_9 load significantly on Factor 4 (F4) named "Flexibility and production potential."

In the second step, the clustering TwoStep technique is applied to the scores obtained by each observation in the data set verified the existence of a structural homogeneity among the organic farms in the sample. This kind of clustering analysis permits us to point out three "natural clusters" in the sample with each cluster including organic farms (Figures 2 and 3). These clusters are described in the following paragraphs.

From the sample total of 1,781 balance sheets, Cluster 1 is composed of 354 observations of which 56 are organic. This cluster mainly includes mid-size farms in the hilly area with vineyards and dairy production. Household members are the main source of labor, and most of the farm's operations are carried out using their own equipment shown by the low level of mean factor score 3 compared to the other clusters.

Table 2. Factor analysis: results of VARIMAX rotation

Variables	F1 Assets intensity	F2 Labor intensity	F3 De-structuration	F4 Flexibility and production potential	Communality
X_8 Debiti	.879				.779
X_5 Uulus	.838				.711
X_{10} Capese	.646				.445
X_4 Uuluf		.785			.627
X_3 Ucv		.738			.548
X_1 Sau		−.602		.387	.523
X_6 Macchi	.303	.556		.266	.474
X_{11} ALT			−.766		.596
X_7 Nolegg			.739		.551
X_2 Sauirb				.696	.514
X_9 Capfop				−.561	.340
Eigenvalue	1.988	1.866	1.195	1.059	6.108
% of variance	18.070	16.967	10.867	9.624	55.528

Source: Authors' calculation on survey data.

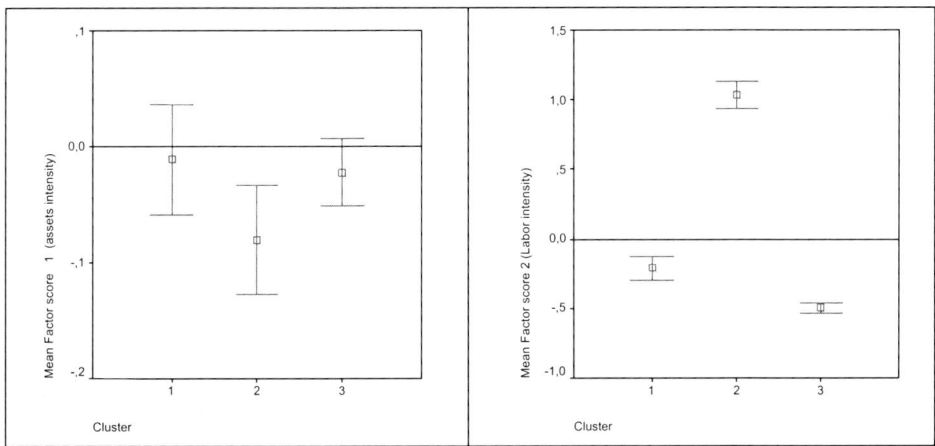

Figure 2. Descriptive statistics of factor scores 1 and 2: plot of the mean and its 95% confidence interval
Source: Authors' calculation on survey data.

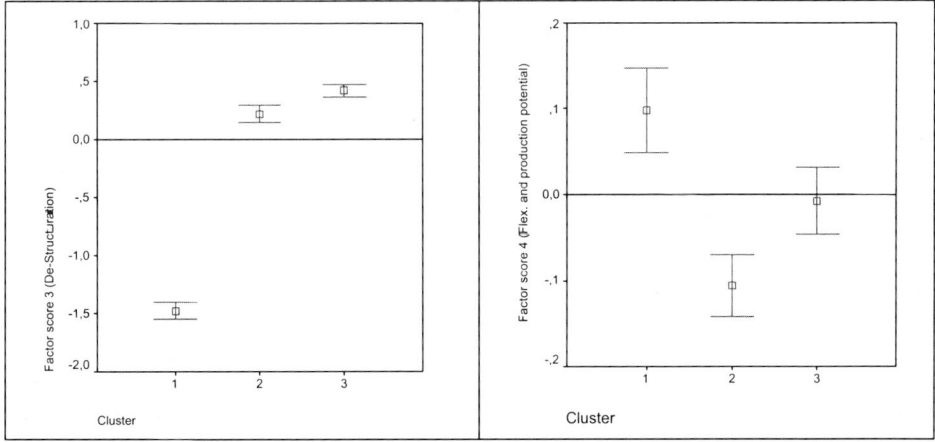

Figure 3. Descriptive statistics of factor scores 3 and 4: plot of the mean and its 95% confidence interval
Source: Authors' calculation on survey data.

The remaining two clusters are located mainly in the plain area. Cluster 2 is composed of 506 balance sheets of small farms, of which 17 are organic. These farms are mainly operated directly by the farmer and are oriented to fruit growing, activity characterized by a high intensity of labor as shown by the high level of mean factor score 2.

In cluster 3, mid-size farms specialized in arable crops prevail, and outsourcing for the farm's operations is quite common. Just 10 out of a total of 921 observations grouped in this cluster are organic.

In the third step, logistic regression is unable to explain and foresee the choice of the farmers when only structural factors are used as explanatory variables, That is, the low level of the pseudo-R square index, R_L^2, in all three groups lets us reject the H_1 hypothesis that structural variables are related to the adoption of organic farming (Table 3). This rejection highlights the need of further information on other types of factors, especially those more closely linked to personal characteristics of farmers. A slightly higher level of the pseudo-R square index in cluster 1, suggests that structural variables may help explain farmers' choices if other variables were to be included also. The adoption of organic farming seems to be more linked to (1) farming systems strictly related to a household's control of the enterprising activity (factor 3 $\beta^{***} = -1.70$) and (2) less labor intense activities (factor 2 $\beta^{*} = -0.55$).

In the fourth step, the multiple regression analysis of per area income results and several structural and diverse variables rejects the H_2 hypothesis of a correlation between the choice of adopting the organic farming and the income level. The t-values associated to the β coefficients of the dummy variables distinguishing the organic and non-organic farms within each year show that the effects of this farm feature are statistically not different from zero for all 3 clusters and years considered. We interpret this statistical insignificance to mean that the choice to adopt organic farming is not relevant for the farm results as measured by income. That is, both organic and non-organic farmers have succeeded in obtaining similar results from their choices and actions.

Table 3. Logistic regression results explaining choice of organic production techniques, by cluster

	B	S.E.	Wald	df	Sig.		
						Cluster 1	
Intercept	−5.01	0.55	81.80	1.00	0.00	−2LL test	134.7
F1	0.09	0.57	0.02	1.00	0.88	Pseudo R^2	0.115
F2	0.50	0.31	2.67	1.00	0.10	Hosmer and Lemeshow (Sig.)	0.371
F3	0.25	0.35	0.53	1.00	0.47	% obs. correctly classified	
F4	−4.35	1.16	13.98	1.00	0.00	− BIO	0.00
						− CONV	99.8
						Cluster 2	
	B	S.E.	Wald	df	Sig.		
Intercept	−4.55	0.49	84.68	1.00	0.00	−2LL test	257.89
F1	0.67	0.44	2.36	1.00	0.12	Pseudo R^2	0.165
F2	−0.55	0.32	3.04	1.00	0.08	Hosmer and Lemeshow (Sig.)	0.739
F3	−1.70	0.26	42.55	1.00	0.00	% obs. correctly classified	
F4	−0.69	0.61	1.27	1.00	0.26	− BIO	19.6
						− CONV	97.6
						Cluster 3	
	B	S.E.	Wald	df	Sig.		
Intercept	−3.30	0.56	35.06	1.00	0.00	−2LL test	101.6
F1	0.77	0.57	1.80	1.00	0.18	Pseudo R^2	0.071
F2	1.91	1.14	2.80	1.00	0.09	Hosmer and Lemeshow (Sig.)	0.059
F3	−1.46	0.84	3.06	1.00	0.08	% obs. correctly classified	
F4	1.47	1.10	1.80	1.00	0.18	− BIO	0.00
						− CONV	100.00

Source: Authors' calculation on survey data.

The high level of model fit (adjusted R-square higher than 70%) can be explained by the high significance of farm's sales (Rictot), a basic factor for achieving a positive result as measured by income, the independent variable in this regression. Since our aim was to highlight the effect of other factors, we did not include the cost of production in order to avoid a perfect fit and thus a trivial result of the analysis. The choice of the activities and some structural factors are significant and influence the farm's results-mainly for cluster 3.

4.2. Emilia-Romagna comparisons

In the fifth and separate step in this paper, the accounting data for organic and conventional farms from 2000 to 2003 show the organic farms in cluster 1 to differ from those in clusters 2 and 3 both in the lower absolute level of and the positive trend (+2.5%) in gross cash farms income (i.e., sales; Table 4). The intermediate costs of the organic farms in cluster 1 have decreased strongly (−15.7%), thus, providing a positive trend in net income (+21.0%) although at the lowest average level. The organic farms in clusters 2 and 3 have higher levels of net income, because of their higher levels of sales. These farms have the opportunity to benefit both from the public subsidies to organic farming and from production levels that do not differ significantly from the conventional ones. During 2000–03, the intermediate costs for organic farms in cluster 3 have tended to shrink considerably faster than their sales, and their increasing rate of net income is the highest (+36.7%).

Table 4. Multiple regression analysis: results

	Cluster 1		Cluster 2		Cluster 3	
Variables	Std. Beta Coeff.	t	Std. Beta Coeff.	t	Std. Beta Coeff.	t
BIO_2000	−0.18	−0.64	0.03	0.21	−0.01	−1.19
BIO_2001	−0.04	−1.17	−0.01	−0.92	0.00	0.38
BIO_2002	−0.04	−1.30	−0.01	−0.84	0.00	0.20
BIO_2003	−0.01	−0.34	−0.01	−0.51	0.00	−0.23
Rictot	0.92***	27.79	1.09***	54.23	1.08***	131.37
SEMINAT[a]	0.05	1.59	0.04	2.45		
ORTFRUTT	0.04	1.36	0.02	1.07	0.03***	4.40
VITE[b]					−0.36***	−4.46
BOVINI	0.06*	1.70	0.02	1.34	−0.07***	−8.69
SUINI	0.04	1.24	−0.18***	−10.62	−0.18***	−21.71
P_COLT	0.08	2.52	0.03	1.88	−0.01	−0.87
COLT_BES	0.03	0.97	0.01	0.73	−0.01	−1.77
Factor 1	−0.12***	−3.55	−0.12***	−6.35	−0.18***	−21.16
Factor 2	−0.13***	−3.65	−0.04	−2.31	−0.12***	−10.09
Factor 3	−0.02	−0.68	−0.04	−2.22	−0.04***	−4.88
Factor 4	−0.05	−1.35	−0.03	−1.43	−0.07***	−5.70
Adj. R-square	0.73		0.89		0.95	
df1	15		15		15	
df2	338		490		915	
Sig. var. of F	0.00		0.00		0.00	

* $p<0.10$, ** $p<0.05$, *** $p<0.01$
(a) variable excluded during the analysis of cluster 3
(b) variables excluded during the analysis of clusters 1 and 2.

In cluster 2, composed mainly of fruit growing farms in the plain areas, the sales and net income levels are higher than the other clusters but this performance must be evaluated with prudence, considering the negative trends in both sales and net income.

Comparing the economic results of organic and conventional farms, we can find seemingly large differences between average values but the significance of these differences is often quite small, because the variability within the sample is very strong (Table 5a,b). In all the clusters the most significant differences may be found in debt payments. Specifically, in cluster 3 the amount of debt payments in the organic farms is double than in conventional farms, situation that highlight a well-established structure. A significant difference in intermediate costs can be found only in cluster 3, where the level for organic farms is almost twice than that for conventional farms. The differences in other categories and clusters are not significant. Net income, expressed as a percent of sales, is quite similar between organic and conventional farms within a cluster, even though they do differ between clusters. So, in most cases we may state that the organic and conventional farms obtain quite similar results that are consistent with the findings in the previous section.

Table 5a. Economic results of the organic and conventional farms within the Emilia-Romagna sample 2000–2003

	Organic		Conventional		F value for the average	Sig.
	Average 2000–2003	Annual change	Average 2000–2003	Annual change		
	– Cluster 1 –					
Gross cash farm income	4,911.19	2.5	4,932.63	–16.6	0.04	0.84
Intermediate costs	2,050.98	–15.7	1,959.72	–6.0	0.08	0.78
Gross added value	2,860.21	25.2	2,972.91	–23.1	0.01	0.92
– Debt payments	685.80	2.0	559.53	21.3	1.38	0.24
Net added value	2,174.41	35.4	2,413.38	–32.3	0.09	0.77
– Wages and taxes	510.54	96.3	530.78	–4.9	0.87	0.35
Operating gross income	1,663.87	25.3	1,882.60	–40.8	0.24	0.62
– Interests and rents	108.47		108.56	7.9	0.06	0.80
Net Income	1,555.40	21.0	1,774.04	–44.2	0.26	0.91
Average age of operators	47		51			

Table 5b. Economic results of the organic and conventional farms within the Emilia-Romagna sample 2000–2003

	Organic		Conventional		F value for the average	Sig.
	Average 2000–2003	Annual change	Average 2000–2003	Annual change		
		– Cluster 2 –				
Gross cash farm income	8,741.62	–24.1	8,532.00	–13.1	0.15	0.70
Intermediate costs	2,087.96	–8.9	2,877.02	–16.3	0.24	0.62
Gross added value	6,653.66	–27.2	5,654.98	–11.5	0.06	0.80
– Debt payments	1,799.77	–1.4	996.40	19.1	3.29	0.07
Net added value	4,853.89	–37.6	4,658.58	–16.9	0.26	0.61
– Wages and taxes	1,086.66	14.3	979.33	–0.20	0.23	0.63
Operating gross income	3,767.23	–53.1	3,679.25	–21.3	0.34	0.56
– Interests and rents	203.18		97.29	33.9	2.12	0.15
Net Income	3,564.05	–57.7	3,581.96	–22.5	0.37	0.54
Average age of operators	47		53			
		– Cluster 3 –				
Gross cash farm income	6,439.56	–16.1	4,048.96	–12.3	0.73	0.39
Intermediate costs	2,653.32	–26.3	1,449.42	–9.7	1.37	0.24
Gross added value	3,786.24	–5.9	2,599.53	–13.7	0.35	0.55
– Debt payments	799.70	–35.1	363.35	14.7	11.66	0.00
Net added value	2,986.54	5.6	2,236.19	–17.3	0.18	0.67
– Wages and taxes	744.94	–39.2	556.79	3.3	0.33	0.57
Operating gross income	2,241.59	29.3	1,679.39	–22.9	0.13	0.72
– Interests and rents	79.03	–50.2	137.40	–2.3	0.72	0.40
Net Income	2,162.56	36.7	1,542.00	–24.2	0.16	0.68
Average age of operators	43		55			

Compared to conventional farms, organic farms seem more able to maintain the capacity to create new wealth over time (see Figure 4, above the line describing wages and taxes). This may be attributed to a more stable system. These results are consistent with the opinions of other authors (e.g. Santucci, 2002b) who maintain that the main motivation pushing farmers to convert their farm to organic agriculture is their aim to preserve the farm's revenues. The level of subsidies for organic agriculture probably plays an important role in this situation. Thus, an interesting issue may be to forecast the reaction to a variation of the level of public subsidy and a higher dependence on the market.

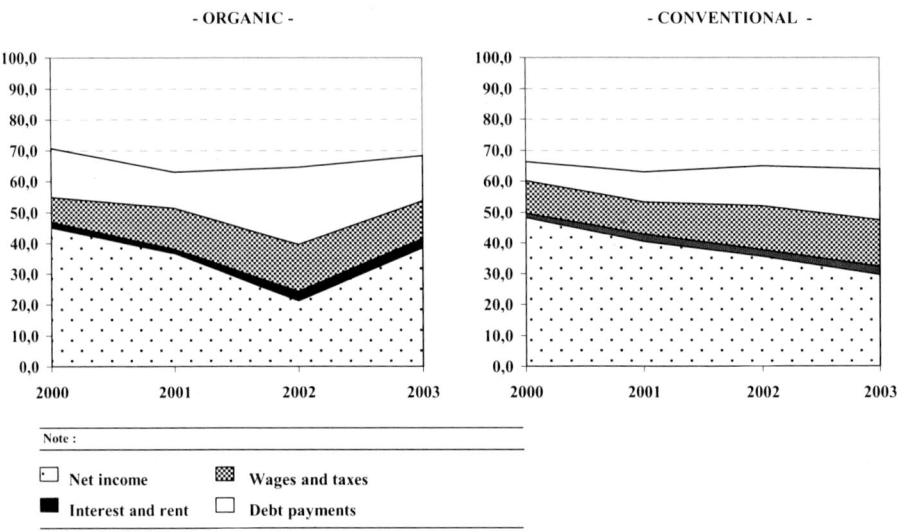

Figure 4. Distribution of added value in organic and conventional farms (percent values)

4.3. Minnesota comparisons

At first look, the organic farms in the Minnesota data have an average net farm income that is lower than the conventional farms, but the difference between the level of net farm income is much less than the difference between the levels of gross farm income (Table 6). The average net farm income for the organic farms over the four years was USD 27,755 per farm. For the conventional farms during the same time period, the average net farm income was USD 35,276. For the organic farms, their net farm income was 19% of their gross cash farm income compared to 16% for the conventional farms.

Average rates of return to assets (ROA) and equity (ROE) show a mixed story. The 4-year average ROA was 5.9% for the organic farms and 5.8% for the conventional farms. The rank was reversed for ROE; the organic farms had an average ROE of 4.8% and 5.1% for conventional farms.

Financial efficiency, as measured by operating profit margin and asset turnover rate, appears to be similar when comparing the 4-year average. But, the annual averages for the two groups do vary considerably and do not follow a similar pattern of deterioration and improvement.

Table 6. Average financial performance and other characteristics of selected organic and conventional farms in Minnesota

Income Summary	Unit	Organic	Conventional
Number of farms	count	42	2897
Gross cash farm income	USD/farm	143,211	225,215
Total cash farm expense	USD/farm	−105,660	−180,766
Inventory change	USD/farm	−548	+7,610
Depreciation & capital adjustments	USD/farm	−9,248	−16,782
Net farm income	USD/farm	27755	35276
Profitability			
Labor and management earnings	USD/farm	16,759	20,035
Rate of return on assets	%	5.9	5.8
Rate of return on equity	%	4.8	5.1
Operating profit margin	%	14.7	14.5
Asset turnover rate	%	39.9	40
Solvency (market)			
Number of sole proprietors	count	40	2,662
Ending farm assets	USD/farm	454,032	707,293
Ending farm liabilities	USD/farm	187,787	358,553
Ending farm net worth	USD/farm	266,245	348,740
Ending farm debt to asset ratio	%	41	51
Crop Acres			
Total acres owned	acres/farm	186	235
Total crop acres	acres/farm	374	547
Total crop acres owned	acres/farm	142	188
Total crop acres cash rented	acres/farm	211	338
Total crop acres share rented	acres/farm	21	21
Labor Analysis			
Number of farms	count	40	2,833
Average age of operators	years	46.7	44.6
Total unpaid labor hours	hours/farm	2,189	2,344
Total hired labor hours	hours/farm	448	478
Total labor hours per farm	hours/farm	2,638	2,822
Unpaid hours per operator	hours/oper.	2,037	2,136
Value of farm production/hour	USD/hour	48.25	74.51
Net farm income/unpaid hour	USD/hour	12.13	14.88
Total labor hours per crop acre	hours/acre	7.1	5.2

The average financial condition of the farms also varies. The conventional farms are larger in terms of the value of assets. The 4-year average value of total farm assets is USD 707,293 for the conventional farms and USD 454,032 for the organic farms with assets valued at market. The organic farms are carrying a smaller debt load, however, relative to their asset value. With assets valued at market, the 4-year average farm debt-to-asset ratio was 51% for the conventional farms and 41% for the organic farms. Annual variation in the debt-to-asset ratio is not very great, and the difference in the relative debt load is also stable over this time period.

Conventional farms are larger in terms of crop acreage: 547 crop acres (221 hectares) versus 374 acres (151 hectares) for organic farms. The absolute number of acres rented is greater for conventional farms, but rented land as a percentage of total cropland is only slightly larger for conventional farms: 66% compared to 62%.

The amount of labor used on the farm also varied between the two groups of farms. The conventional farms used more total hours per farm while the organic farms used more labor per crop acre. The conventional farms had a higher level of net farm income per hour of unpaid labor.

In general, this analysis seems to support the widely held view of organic farms in Minnesota: on average, the organic farms in this data-set, tend to be smaller, carried less debt, had lower crop yields, and produced less income per farm and per hour compared to conventional farms. However, since individual data were not accessible, the significance of these differences cannot be tested.

5. FINAL REMARKS

In this analysis and with the available data, we were unable to show that income differences were due to adopting organic production methods. We rejected both the hypothesis that structural variables are related to the adoption of organic farming and the hypothesis that a direct relationship exists between income level and organic farming adoption. Thus, income seems to be more related to individual farmers' abilities to manage resources and processes, not organic versus non-organic. However, since our data did not include some possibly important variables, future research should utilize an integrated dataset to analyze economic and financial performance as a function of farm structure, management practices, and personal and household characteristics.

6. REFERENCES

Jarvis, L.S., 1981, Predicting the diffusion of improved pastures in Uruguay, *American Journal of Agricultural Economics*, **63**: 495–502.
Kaiser, H.F., 1974, An index of factorial simplicity, *Psychometrika*, **39**: 31–36.
Menard, S., 2002, Applied logistic regression analysis, in: *Quantitative application in the social sciences*, Sage University Paper, Thousand Oaks, California.
Neupane, R.P., Sharma, K.R., and Thapa, G.B., 2002, Adoption of agroforestry in the hills of Nepal: a logistic regression analysis, *Agricultural System*, **72**: 177–196.
Prochaska, J.O., Di Clemente, C., and Norcross, J.C., 1992, Search of How People Change: Application to Addictive Behaviours, *American Psychologist*, **47**: 1102–1114.
Rogers, E.M., ed., 1995, *Diffusion of innovation*, The Free Press, New York.
Santucci, F.M., and Chiorri, M. 1996, *Economia delle produzioni biologiche: il caso dell'Umbria*, Quaderno 19, Istituto di Economia e Politica Agraria, Perugia.
Santucci, F.M., 2002a, Limiti e necessità della comparazione tra biologico e convenzionale, in: *L'agricoltura biologica in Italia: metodologie di analisi e risultati dell'utilizzo dei dati RICA*, INEA—I metodi RICA, Roma, pp. 15–21.
Santucci, F.M., 2002b, Convenienze micro-economiche dell'agricoltura biologica, in: *Problematiche dell'agricoltura italiana. Scenari possibili*—Agricoltura biologica, Accademia Nazionale di Agricoltura—Consiglio Nazionale delle Ricerche, Bologna, pp. 125–185.
Sterns, J.A., and Peterson, H.C., 1996, *The propensity to enter and exit export markets*, Staff Paper 96-30, Department of Agricultural Economic, East Lansing, Michigan.
Young, K.D., and Shumway, C.R. 1991, Cow-calf producers' perceived profit maximization objective: a logit analysis, *Southern Journal of Agricultural Economics*, **23**: 129–136.

Zanoli, R., Gambelli, D., and Fiorani, S., 2002, La comparazione economica tra aziende biologiche e convenzionali: aspetti metodologici e strumenti operativi, in: *L'agricoltura biologica in Italia: metodologie di analisi e risultati dell'utilizzo dei dati RICA*, INEA—I metodi RICA, Roma, pp. 25–40.

SITUATION AND PERSPECTIVES OF ORGANIC MEAT IN ITALY[*]
The experience of a small group of firms located in the Veneto Region

Luigi Galletto[**]

SUMMARY

Although organic meat consumption has recently shown a great percentage growth in some EU countries, different problems have risen in order to supply the markets with adequate levels of organic meat, especially in countries like Italy where conventional meat production is based mostly on very intensive methods.

The first part of this paper is a review of the production and market situation of organic meat in Italy. The second is a preliminary attempt to examine the structural features and the marketing strategies of a pioneer group of firms which have begun to produce and/or to process and market different types of organic meat in the Veneto Region, the most important area for the production of conventional meat in Italy.

Besides some satisfactory results, the research has shown many difficulties related either to the costs or to the prices faced by organic farms. The costs are mainly due to the wide gap between organic and conventional production techniques. Prices are typical of the building up of a new niche market. For now, these difficulties seem to prevent a rapid expansion of domestic supply of organic meat in Veneto, although specific research programs could overcome some of them.

1. ORGANIC MEAT SITUATION IN ITALY

The regulation context has not been very encouraging for organic meat production. Although in 1991, EU Regulation 2092 established a series of technical norms regarding exclusively plant organic products. It was only in 1999 that EU Regulation 804 regulated

---

[*] Research funded by the Venetian agricultural Agency, "Veneto Agricoltura."
[**] Dept. TeSAF, University of Padova, Agripolis Via Romea, 35020 LEGNARO (PD) Italy; e-mail: luigi.galletto@unipd.it.

animal organic production. Moreover, the agro-environmental measure provided by EU Regulation 2078/92 granted per hectare subsidy to organic agricultural cultivations, but no aid was allocated for organic breeding. Also Regulation 1251/99, which gives financial support to organic agriculture for a five-year period by means of the regional Rural Development Plans (RDP), has moved on the Reg. 2078/92 path, and targets aid only to organic crop.

Despite this situation, consumption of organic meat has increased strongly in EU countries and in Italy during the last few years, due to recent food scandals that hit the food industry. In particular, the return of BSE disease in 2000 has again raised the interest for organic meat from consumers who are more aware of the problems related to healthy nutrition. Only now, is organic meat beginning to be sold in more and more outlets, an encouraging sign that it may leave the micro-niche in which it has remained up to now.

1.1. Production

At the end of 2000, the Italian organic certified area (including the one in conversion) was about 1,070,000 hectares, approximately 7% of the Italian total usable agricultural land. In comparison with the previous year, a 12.2% increase has taken place, especially from larger farms, although, this rate increase was lower than in the previous five years (approximately 20% per year). The highest increase happened between 1997 and 1999. This period incorporated the widest implementation of the EU Regulation 2078/92, which supported organic agriculture methods.

According to the latest available data, there were 468 organic livestock farms in Italy at the end of 1999. More than 60% of these farms were located in northern Italy. In particular 16% were found in the Province of Trento and 11.8% in Veneto, with an incidence on total organic agricultural farms respectively of 22.2% and 7.4% (Lunati, 2001b).

In 2000, 1,423 animal breeding farms requested to join the national organic certification system. A large percent (47%) were meat producers, 28% specialized in milk production and the remaining quarter were mixed farms (Lunati, 2001a).

From the production viewpoint, two main organic farming techniques are present: a more intensive one, which implies modest changes in the existing structures, and a more extensive one, which mostly adheres to the principles of the organic animal production regulation. The intensive technique appears to be more focused on the mass market, and the extensive technique more focused on niche markets. Operators using the more extensive one may not obtain immediate benefits from the exploitation of mass market, but they could achieve, in the medium term, a greater flexibility in the productive choices and be less exposed to competitive pressures. This is due to lower fixed costs, especially if they succeed in connecting organic production with typical elements (Di Marco, 2001).

From an economic viewpoint, recent research by Cozzi *et al.* (2001) has found the production cost per head of organic beef meat to be considerably higher than meat produced conventionally. The greatest difference is in feeding expenses (about 30% higher) which is partly due to the absence of a true market for organic feedings stuffs. Also, organic meat producers had higher investment depreciation and an additional cost for control and certification. Health care expenses, however, are lower. Revenues, which increase either by greater slaughter weight or by higher price, provide a 50% increment in the per head net income for organic producers. However, if the budget is based on the production cycle span, such advantage vanishes. Net income is 25% lower for organic producers if the budget is carried out over a five-year period for the overall business, due to limitations on stocking. Because of this current lack of competitiveness, a massive organic conversion of Italian meat production does not seem reasonable, not only for beef but also for other meats where high production costs are reported. This happens especially in chicken meat production (Didero and Troiani, 2001). Other authors (Rossi and Gastaldo, 2001) have shown that the investment costs per head for structures used in the organic pork meat production are not higher than those in the conventional breeding; however appraisal is more difficult if you need to adapt already existing structures rather than to build new ones.

1.2. Market

Although there are various technical-economic problems in the production because certified operators who follow breeding production in the chain are scarce, the distribution of organic meat is also very incomplete. Moreover, transactions along the chain are still at minimal exchange volumes. The market share is somewhat uncertain (from 0.5% to 2% of the total value of the meat sold) and most of the meat is imported, given the huge domestic production deficit (Baron, 2000).

Organic meat is available in approximately 120 sale outlets, located mostly in northern Italy, where 70% of the national consumption occurs. This happens for two reasons: (1) a growing interest from operators in the large scale retail (LSR) distribution system where we can estimate that, in some cases, meat accounts already for 4–5% of total organic food sales and (2) interest from a network of retail shops, who specialize in supplying organic meat.

Both traditional butcheries and the LSR have shown a remarkable interest for organic meat, as long as dealing with it does not involve excessive logistic and managerial difficulties (Didero and Troiani, 2001). A recent survey, which was carried out in Piedmont, has examined the opinion of chain operators who follow the breeding phase (Marengo et al., 2000) and has drawn some interesting indications. With reference to slaughterhouses, there are no particular difficulties provided that processing organic meat separately from conventional meat is a feasible operation. This method involves low additional costs, from 1% for cattle meat to 5% for chicken meat. At the small conventional retail level (butcher shops), research has found a remarkable interest for organic meat, in particular for beef (53%) and chicken (27%). Moreover, within these operators, preference for fresh meat has appeared to be greater (72%) than processed meat (22%). Eighty-seven percent of operators think a 20% premium price over the conventional meat to be reasonable and more than half of them would be willing to accept a 30% maximum increment. Proper information, high quality and affordable prices are ranked as the key factors needed to boost organic meat in this trade channel. As far as organic specialized shops, although they admit a willingness to pay a maximum 50% premium price for organic meat among their customers, they have revealed a lower propensity for the organic meat sale than the butcheries. There was a little higher interest for the poultry meat in comparison with beef and a preference for preserved meat versus fresh meat. LSR operators have shown enough interest for organic meat, in particular for poultry and beef, but they believe it is too expensive for their usual customers, provided that 30% would be the maximum acceptable premium price in some cases. Although, 15% would be a more adequate premium for the LSR operators.

Finally, there is a substantial readiness to deal with organic meat, based perhaps on an optimistic view of its market perspectives. Although, the interests of the various distribution channels seem to be differentiated. The strengthening of the organic meat market appears to be mainly associated with its spreading among the LSR sale outlets and the butcheries. But, at least in the short period, the requirements of the LSR channel do not appear to be met by domestic supply because supermarkets are not ready to deal with small quantities and in a discontinuous way. During an intermediate period, the specialized shops can play a certain role, due to the difficulties arising from managing the two kinds of meat in the same outlet, which would limit the organic meat option to a small number of butcheries.

Direct selling is also not easy. In this regard, we can draw some insights from the experience of some producers in the Marche Region who have undertaken the production of organic meat as an opportunity for better rewarding their labor (Ansaloni and Bellavia, 2001). While those who kept selling live cattle have not achieved any premium price. Those who have begun to pack a variety of beef cuts at the farm level have obtained a price which is considered satisfactory by half of them. This is due to the fact that most of their customers were people living in rural areas, for which price is still the determining factor in deciding the purchase of food products. Therefore, in these circumstances, the association of organic producers in cooperative firms appears to be the best option in order to provide supply concentration and penetrate new markets.

Similar to other organic products, organic meat requires particular attention to price policies and strong promotion linking information about its features to the consumer. On average, a 50% price premium is currently applied to organic meat, whereas for dairy products it is considerably lower (19%). A reasonable price should not be higher than 30–40% in relation to conventional meat (and 20% for high quality cuts). Some supermarkets gradually approached this level of price premium at the end of 2000 (Didero, 2001a).

1.3. Perspectives

In the short-term, Databank (2001) forecasts a 50% increase in organic food sales in LSR with meat playing a role in the increase. The organic meat segment has a competitive advantage, thanks to the recent food alarms and to consumers' requests for food safety.

Although meat does not play the primary role within organic products, since it is forecasted to only average a 4–5% share in the total meat consumption through the next decade (Menghi et al., 2001). The long run target, even though small compared to other organic products (10% or more), relies on organic meat's differentiation in relation to tenderness, taste characteristics and price from conventional meat. However, the low level of standardization of organic meat (especially for color and tenderness) can be a barrier to purchase even for those who initially favor its consumption (Didero and Troiani, 2001).

Current national supply is not able to meet the demand for organic meat. This is particularly important for the LSR which has to buy larger quantities than the other channels and is forced to rely on imports from Austria, Germany and the Netherlands. For this reason, the main problem, which now affects all organic meat operators, is how to meet the market demand. The organization of the domestic supply, which is closely related to the organic breeding farm profitability, plays a crucial role in supplying the market in a sufficient and continuous way, without relying exclusively on imports (Marengo *et al.*, 2000). This problem is also limiting the increasing use of organic meat in school catering, even though city administrations can pay higher prices than other consumers (Didero, 2001b).

In the case of beef meat, the best perspective seems to incorporate both the Italians traditional breeds, as Chianina and Marchigiana, and other French breeds. The enhancement of indigenous breeds, which are endowed with remarkable robustness, represents one of the most important aspects of organic animal breeding oriented to meat production. In fact, the local characteristics of the raised breed can find an added value multiplier due to the safety of the production methods and the reduced environmental impact. For Italian breeders, organic meat is an opportunity to be developed especially where it is easier to use extensive breeding techniques and, therefore, may be a way to revitalize some mountain and hill areas where intensive breeding is too expensive. However, growth in organic meat appears less encouraging in the Po Valley due to the use of intensive methods (Baron, 2000).

Concerns have been raised about organic pig breeding (Menghi *et al.*, 2001). Organic pig breeding appears to be hard to manage without the use of some medicines and also due to the frequent heterogeneity in traded meat. However, the scarcity of some organic herbs needed in processing gives the whole meat, which is preserved by using only salt, a better market opportunity than meat obtained through a milling process.

2. OBSERVATIONS FROM A SMALL SAMPLE OF VENETIAN FIRMS DEALING WITH ORGANIC MEAT

Similar to the Italian meat industry the Venetian organic industry has started later than other organic product sectors. This was due to the lack of a clear legal reference frame and low market demand. Recently organic meat was an unknown product to consumers and also to agricultural producers. In the 1990s, organic animal production in Veneto was very rare and limited to people who tried to reinstitute traditional breeding methods, mostly in poultry breeding. Producers tried to raise the animals according to the rules proposed by the organic agriculture associations. These associations, lacking specific regulation, certified animal production as organic through the use of techniques recognized by the International Federation of Organic Agriculture Movements (IFOAM).

The increase in meat demand has recently widened operators' interest in this region, thus leaving room for more research to explore this newborn industry.

The current survey was done with eight enterprises. Two belong exclusively to the organic meat processing and trading phases, while the others six are also involved in or solely include the meat production phase. Within these six, one is specialized in sheep meat, one in pork meat, two in poultry and two in beef meat. The sample size is due to the limited presence of firms involved in organic meat within the regional territory. In fact, very few farms are already fully converted to the organic meat production techniques and totally to the regimen. The organic meat-processing part of the chain is still at an embryonic stage.

The eight enterprises are scattered as follows:

- in the Venice Province: the smaller size processing firm;
- in the Verona Province: the lager size processing firm and the smaller size cattle breeding farm;
- in the Padua Province: the larger size cattle breeding farm, the pig breeding farm, the sheep breeding farm and the smaller size poultry breeding farm;
- in the Treviso Province: the larger size poultry breeding farm.

The data presented here has been obtained by means of direct interviews of farm entrepreneurs and firm managers. They were carried out in the second half of 2001, with the aid of an *ad hoc* questionnaire.

2.1. Some structural characteristics of the organic meat breeding farms

With the exception of the larger cattle breeding farm, which has a separate sale management, the five breeding farms are all individual firms. The breeding farm dimension range is somewhat wide; the agricultural usable surface varies from 5 hectares of the larger size poultry breeding farm to 220 hectares of the farm with the sheep flock (Table 1). In the larger cattle breeding farm and in the pig farm, the surface in conversion is still wider than that already converted to organic agriculture. Nearly a third of the surface of the latter is still used for conventional agriculture. In the larger poultry farm, three vineyard hectares are cultivated conventionally. Originally this was a wine farm which has undertaken organic poultry by using two vineyard hectares as pasture place for the reared fowls. The additional feed is supplied by three rented hectares of arable land used to grow cereals; the remaining feed deficit is satisfied by buying organic feed. In the second poultry breeding farm, organic meat production still has a marginal function. Its context is not only of a multi-crop organic farm (an orchard and a vineyard are present beside the arable land surface) but also a multifunctional one, provided that the farmer cultivates specific crops for feeding game. Only the sheep farm represents an exclusive breeding farm, where 20 hectares are hayed and the remaining is pastureland. The cattle breeding farms have also some orchard and vineyard surfaces, although, in the case of the larger farm, the size of the vineyard is rather small when compared with the size of the arable land.

Table 1. Surface and replacement value of the rural buildings in the sample breeding farms

Breeding farms	Usable agricultural surface (ha)			Replacement value of rural buildings (EUR)
	Total	Organic	In conversion	
Cattle 1 (Padua)	80.0	33.0	47.0	700,000
Cattle 2 (Verona)	16.0	16.0		280,000
Sheep (Padua)	220.0	220.0		135,000
Poultry 1 (Padua)	8.0[#]	5.0		125,000
Poultry 2 (Treviso)	11.1	10.6	0.5	350,000
Pig (Padua)	25.0	7.7	9.5	150,000

[#] Of which 3 ha are rented.

The replacement value of rural buildings appears correlated with the farm surface. It seems rather low in the larger poultry breeding and in the sheep farms due to particularly unrefined animal shelters.

The type and number of animals also varies on these six farms (Table 2). The first cattle breeding farm has two barns (each 500 square meters) and a productive cycle longer than the second, weaning at a higher weight. It diversifies its sources of income through supplemental hen production which has a longer cycle than the same species has on both poultry breeding farms. A multiple breeding system also characterizes the farm located in the Padua Province, although turkeys and Guinea fowls give a decidedly lower contribution to the overall revenues than chickens. In the sheep breeding farm, animals are maintained in the pasture without any shelter all year around, but special huts are used in the pre- and post-birth period. Also, pig breeding is based on traditional pasture, for which electrified fencings and plain sheds have been built. In all the other cases, the open stall system is used in compliance with organic breeding rules. Both the cattle breeding farms and the larger poultry breeding farm sell slaughtered animals, while the others deliver live animals.

2.2. The managerial resource

Also in the organic Venetian breeding farms, like most of the conventional farms of the Region, entrepreneurial and family labor largely prevails over the directly hired labor or machinery contractors. Family labor is always intensely involved in the usual breeding activity, and occasionally the farmer commits other workers in crop cultivation.

In the interviewed entrepreneurs' opinions, the professional ability required in various activities connected with the organic breeding production process is mostly upper-middle, especially for health care and pregnancy induction concerns. Slaughtering and carcass processing are always done by other firms, but the entrepreneur of the cattle breeding farm located in the Padua Province expressed the intention to build a farm slaughterhouse.

Most of the breeding farms have been in the breeding business for a long time and converted their traditional methods to organic methods. Only for one farm is breeding a

Table 2. Livestock characteristics of the sample breeding farms

Breeding farms	Animal type	No. of heads	Initial weight (Kg)	Final weight (Kg)	Cycle span (Months)
Cattle 1	Beefs	250	150	650	16–18
	Cows	50	550	550	
	Hens	500	Chicks	2.5	6–7
Cattle 2	Beefs	80	250	550	8
Sheep	Lambs	1000	6–7	18–20	2
	Fat lambs	*	6–7	50–60	6–8
Poultry 1	Chickens	3000	Chicks	3	4
	Turkeys	400	Chicks	4.5	7
	Guinea fowls	800	Chicks	1.2	4
Poultry 2	Chickens	250	Chicks	2.5	5
Pig	Fattening pigs	150	25	180	10

* Share of the Lambs in relation to their Easter sales.

new activity. Professional satisfaction was the main incentive for converting to organic production for three farmers out of six. In the other cases, conversion appears to be dictated more by idealistic motivations, as has been true for most types of pioneering organic farms.

In general, we have recorded an extensive demand for professional updating from the organic meat operators. It relates above all to technical and legal aspects, but the sources they use appear quite diversified even within the same firm type. The contribution supplied by agencies which control and certify organic production and processing methods is regarded as particular important by many of them (processing operators included). Most interviewees made reference to the legal issues addressed by the agencies.

However, there is less use of trade and marketing information, which is obtained mostly by customers. Only the non-farm units utilize specialized trade and market publications to such aim. This situation appears in accordance with the prevailing extra-economic motivations for organic breeding pioneers, as previously noted. Persisting in time, however, this could generate serious problems both for the profitability and the sustainability of the organic meat business.

2.3. Some economic elements

Whereas the total revenue for both of the processing enterprises amounts to over 500,000 EUR, the total sale value of all the breeding farms but one is less than 250,000 EUR. Only the cattle breeding farm, with the commercial company, exceeds 500,000 EUR in sales.

The dynamics of revenues in the last three years does not appear to be even across these farms. The smaller cattle breeding farm, which exclusively wholesales, declares it has sustainable stability, while that larger one that practices direct selling and utilizes other trade channels, shows a high growth in the revenues. The high growth is particularly noticeable for the linked trading company.

Sales for the sheep breeding farm (100,000 EUR) and for the small poultry farm (15,000 EUR) are stable. The revenue for the larger poultry breeding farm showed a slight increase during the 1998–2000 period but are expected to double in 2001, which is the first year following the end of the conversion phase. The two processing enterprises have both achieved a striking sales increase. One shows an increase from 1,500,000 EUR in 1999 to 3,500,000 EUR in 2000 and the other has begun its activity in 2000 with 300,000 EUR sales, but it will widely exceed 500,000 EUR in 2001. The pig breeding farm carried out its first sales only in 2000, with a revenue of approximately 20,000 EUR.

Also in a small number of cases, prices show a remarkable variability within the same kind of breeding, in relation to the different sale channels utilized.

As for as the meat processors' price, we comment on the information from only one beef firm. On this farm the gross sale price is, on average, 25% higher than the conventional meat price, with the exception of the organic specialized shops, where the processor obtains an approximately 50% increment for pre-packed meat. The retail price increase in the outlets supplied by the firm ranges from 25% in LSR sale points to 60–70% in the organic specialized shops, and it is around 35% for consumers in the other channels.

The survey has also provided some interesting information on cost elements. The two cattle breeding farms, the small poultry and the pig farm reported some values regarding fodder crops. In the two cattle breeding farms (which use exclusively their own manure as fertilizer) cash costs per hectare vary between 150 and 180 EUR for soil mechanical operations and between the 200 and 250 EUR for forage harvesting. But the widest difference is in the weed control cost, from 90 to 250 EUR/ha. The organic pork meat producer shows the highest fertilization cost (200 EUR/ha). Organic crop certification costs range from 10 to 20 EUR/ha.

Cost elements relating to the breeding phase appear to be more interesting. In particular, for one cattle breeder the per head purchase cost is approximately 400 EUR, whereas for the other it reaches 750 EUR, depending on the breed type. In these farms, feed expenses vary between 1.45 and 1.70 EUR per head per day, health care expenses between 15 and 21 EUR per head and certification cost of organic breeding are 2 EUR per head in both cases. The per head labor need is much higher in the smaller breeding farm, which requires 1,200 hours per year for stable-related works, against a 1,500 hours requirement from the larger dimensions breeding (250 heads in the fat stock keeping and 50 cows). This difference can also be an index of remarkable scale economies in the organic cattle breeding and it may also be ascribed to constraints connected to the previous farm structure.

In the poultry breeding farm with 250 chickens, which has a self-replacement cycle and an annual labor of only 60 hours, direct per head costs are estimated to be 4.40 EUR, of which 0.50 relate to veterinary care, 0.50 for general expenses and the remaining amount to feeding costs. In the other poultry breeding farm, the annual labor requirement is very different at approximately 1,500 hours. Expenses for such breeding are 0.50 EUR for a hen or Guinea-fowl chick purchase and 2.80 EUR for a turkey poultry purchase. Feed expenses are estimated in approximately 5 EUR per broiler and Guinea fowl and 10 EUR per turkey. To these we have to add a little more than 0.50 EUR per head for general expenses, insurance and veterinary care costs.

The sheep breeding farm shows low costs, which are typical of a pasture-based technique. Labor employment for sheepfold management is approximately 700 hours; whereas what is needed for pig breeding is a little lower (600 hours). In this firm we found the following direct costs per head: 100 EUR for piglet purchase, 200 EUR for feeding, 2.60 EUR for veterinary care.

The purchase prices of organically produced animals to be fattened or to be slaughtered are determined by practicing a percentage mark up on the price of the same animals obtained in the conventional way. It can also be determined by specific supply contracts in which the purchase price of the animal to be bred is linked to that established for its sale. Only one processor gives the supplier a price, which is connected to the breeding production cost.

As far as the processing and marketing costs, the larger poultry breeder calculates approximately 3.60 EUR per head (including the high charge needed to deliver small quantities to the multiple retailers). The smaller cattle breeder estimates that slaughtering, stocking and transportation cost are about 0.50 EUR per Kg of meat. The other cattle breeding farm provides more detailed information, where slaughtering and transportation costs are 90 EUR per head, and cuts selection, packing, labeling and other logistic expenses are around 2.50 EUR per Kg of meat. This is a higher value than that given by one of the two processing firms, namely 1.7 EUR per Kg which includes stocking, shipping, marketing and certification costs.

2.4. Supply and market

Organic meat supply varies according to farm size. Only the larger cattle breeding delivers a meat quantity which is comparable to the quantities shown by the two processors (Table 3). The supply from the two processors appears, however, lower than that of other similar enterprises operating in the conventional meat industry.

There are two different supply strategies in the two cattle breeding farms: the first acquires calves twice a year—in spring and autumn—with at least half of the available places always utilized. The other, instead, makes all the purchases in September and gradually sells the entire production in the spring-summer period. The sheep breeding farm sells all the lambs at Easter time and sells the fat lambs in October.

All the operators consider producing exclusively for the Italian market. Among the channels, the organic specialized shop seems likely to prevail, but among breeders wholesalers are also important. Although direct sales can give more rewarding prices, they remain confined to small amounts and only the smaller poultry breeding uses this channel exclusively.

The larger poultry breeding farm delivers to wholesalers 75% of its turkey meat, while it supplies specialized shops 80% of its chickens and 50% of its Guinea fowls. The commercial company linked to the larger cattle breeding owns a retail outlet in the city of Padua.

The processing firm in the Verona Province supplies wholesalers and has contracted approximately 600 cattle during the year 2000, the same amount of pigs and about 40,000 poultry from breeding farms (hens, chickens, ducks, geese, turkeys and Guinea fowls). Direct sales are carried out in a small chain of shops, which are managed through franchising. Because of the very low availability of Veneto organic meat production, this enterprise has been forced to buy goods on either the national or foreign market. All the organic pork meat and half of the poultry meat come from Italy, the other half of the poultry is imported from France. Eighty percent of organic beef meat comes from the EU market, particularly from Austria; the other 20% is imported from non-EU countries, especially Uruguay. The other processor has succeeded in selling, on the organic meat market, only 20 tons provided that lower quality cuts are sold on the conventional meat market.

Trade relationships with LSR affect only the two processing firms. The one based in the Venice province has begun, only a year ago, to supply supermarkets with organic beef meat. The other has established contacts with supermarkets and/or their chains for three years, given that it has initially privileged distribution to the specialized retailers.

Among the factors that have mainly influenced the product mix choice of the various firms, most of the interviewed operators have indicated the product price as the key element in their choice followed by the production cost. Productive risk, mainly linked with product availability, ranks subsequently, particularly for one of the processing firms and for the larger cattle breeding farm. Marketing risk has been the last factor taken into account, which means that operators expect limited price volatility for their organic meat.

When entrepreneurs have been requested to express their perception on tendencies in the organic meat market they have indicated a growing or stationary trend. Among the

Table 3. Organic meat supply

Firm type	Sales (kg of meat)	Marketing channels	Share %
Cattle breeding farm1 (Padua)	10,000 (chicken)	Restaurants	20
	120,000 (beef)	Specialized shops	80
Cattle breeding farm 2 (Verona)	24,000	Direct sales	25
		Wholesalers	75
Sheep breeding farm (Padua)	35,000	Wholesalers	100
Poultry breeding farm 1 (Treviso)	9,000 (chicken)	Direct sales	20
	1,500 (Guinea fowl)	Wholesalers	20
	1,800 (turkey)	Specialized shops	60
Poultry breeding farm 2 (Padua)	750	Direct sales	100
Pig breeding farm (Padua)	27,000	Direct sales	10
		Processors	90
Meat processing 1 (Verona)	60,000 (pork)	LSR	10
	180,000 (beef)	Conventional butcheries	10
	80,000 (poultry)	Specialized shops	30
		Processors	15
		Catering	30
		Own retail outlets	5
Meat processing 2 (Venice)	25,000 (mixed meat)	Wholesalers	25
		LSR	50
		Specialized shops	25

breeders, only those who direct market beef meat and the pig breeder have assumed a short-term increase of 40% for organic meat sales, whereas the others think that a stationary path is more likely. In contrast the two processing companies have forecast 10% to 40% sale increases in the short term. Consequently, such enterprises believe that similar increases in their supply can find an easy outlet in the organic meat market.

With the exception of the sheep farmer, all the other entrepreneurs declare that organic meat processing introduces substantial differences from the conventional way in relation to meat manufacturing, conditioning and storage. Only one processing firm states that there are also differences also about conservation, whereas no enterprise thinks the trading methods are different.

Half of the interviewed entrepreneurs believe that there is a market for noncertified organic meat, which is mostly sold directly to consumers at the farm level, but their estimations diverge widely on its size. Such a market would be due to the high organic certification cost (especially for the small size farms) and difficulties in complying fully with the organic breeding rules.

2.5. Production technology

Several operations implied that production of the organic cattle meat uses traditional equipment, structure and methods using medium-low technological levels. The technological level of the operations in organic poultry breeding appears to be analogous or a little lower. Choice of technology is also influenced by the size of the breeding herd. No technological improvement is envisaged in sheep breeding.

Rations in the two cattle breeding farms use large amount of forages (hay and corn silage in the smaller one; corn silage, permanent grass hay and dehydrated alfalfa in larger one). The larger farm produces its own concentrate starting from single raw materials (corn, barley and soybean grain and bran), while the smaller farm prefers to utilize a purchased organic concentrate. Feeding in the poultry breeding takes advantage, besides grubbing, of purchased concentrate that comes from abroad. Pig breeding uses, besides *ad libitum* pasture grass, a daily concentrate amount that ranges from 2 to 4 Kg per head, depending on the different phases of the production cycle.

As far as investments carried out during the last five years, the sheep breeder and the smaller poultry breeder stated they had begun organic breeding activity without any significant investment. The other poultry breeder has invested 26,000 EUR in order to undertake organic breeding through developing an open stable system, pasturing and buying a truck. The pig breeder has invested around 15,000 EUR in fencing and other simple equipments. The larger cattle breeder has carried out investments of approximately 255,000 EUR, while the smaller one has invested 37,000 EUR in order to make the structural adjustment required by the organic production method. The slaughterhouse planned by the larger cattle breeder will need an additional investment of 250,000 EUR. The larger processing firm has declared that it has invested more than 500,000 EUR during the last five years, in order to start processing organic meat.

2.6. Marketing strategies

Besides the already noted extra-economic motivations, the reasons for the organic choice are tied substantially to the market demand and the operating margin one hopes to gain by producing organic meat, thanks to the premium price the consumer is willing to pay. Only non-farming firms list enlargement and new product search as motivations. Hence, they are operators who are going to occupy, as pioneers, a space which they think will widen itself. They are also confident in the analogous market growth that has marked the crop organic products in the 1990s. The operators have decided to face the difficulties entailed in the construction of this new market, but hope to receive some economic satisfactions in the long run.

The fact that only the larger cattle breeding has taken advantage of a marketing plan (which was developed by an external advisor) is emblematical of a such primitive phase in the building up of the market. This plan, aimed at expanding sales in the home market, implies the use of all the four marketing mix basic levers (product, price, place and promotion). In both the processing companies, the staff plans marketing strategies and has a higher use of the marketing levers than the breeding farms. In these plans, marketing activity is mainly focused on home market expansion.

Each firm claims meat price stability for every type of product during the year. At the wholesale level, organic meat price is influenced mostly by the production cost, the transaction amount and the type of purchaser.

In the operators' opinions, some key factors appear to be the most important for the improvement of sales flow, with reference to each marketing channel. On-farm direct sale, while it generally results in satisfying prices, shows logistic problems (for one cattle breeding and the pig one) and limited tradable amounts induce the breeder to divert part of the production towards other less rewarding channels. Within these, wholesale is characterized by prices that are lower than the expectations and in some cases by delayed payments. The margin is judged to be low for beef meat that is sold to catering units. The logistic aspect and the limited tradable amounts appear to be the critical points in supplying to specialized shops, where a price beneath the expectations has to be added in the case of the poultry products. Therefore, the marketing channels also have organization problems, which are typical of a building up niche market.

The larger cattle breeding is the only farm to take advantage of several marketing tools at once: a collective brand, a business brand, organic and environmental certification and several advertising techniques, such as farm visits, a booklet, a web site, and sponsoring different public interest events.

Business brand is used also by the pig farm and by the larger poultry farm, while the smaller firm does not take advantage of any communication tool. All farms except the smaller farm utilize the organic production certification. The processing firm located in the Verona Province utilizes two trademarks; one for the specialized shops distribution and the other for being acknowledged by LSR customers. The other processing firm does not adopt its own brand, but takes advantage of a collective brand for organic products which is already famous in the LSR and wholesale level. It also carries out promotions at the outlets where its meat is available.

Among the breeding farms only the cattle farm, which practices direct retail selling, and the pig farm have taken part in events related to organic food, at either the local or regional level. They believe they achieved good or acceptable outcomes regarding the on-farm demand and the brand name. Both processing companies have participated in specialized international promotion events, of which they retain a good, if not optimal, opinion.

The strategies for the improvement of organic meat distribution rely mainly upon putting into practice structural actions within the chain aimed at increasing the supply. Also, channel diversification and promotional actions are considered useful in order to ease the advance of the organic meat chain. This implies that operators perceive the need to increase their bargaining power, which currently appears still rather low.

At last, as far as the strategies for supporting the sale levels, only the sheep breeder does not show any proposal to further promote their own business; he seems to be already satisfied by the current situation. On the other side, the strengthening of quality standards and the name associated with the business brand are strategies, to which both the cattle breeding farms and both the processing firms are tending. Selling price reduction appears to be a necessity only for the processing firms. For three breeders, cutting the cost of meat production is important, too, while two other breeders wish to achieve additional certifications in order to better guarantee their product goodness. The use of a well-known certification brand is a need recognized only by the pig breeder, and only one breeder trusts in a greater development of the direct sale channel.

3. CONCLUDING REMARKS

Although we need a noticeable caution in drawing conclusions, due to a heterogeneous and numerically very small sample, our survey seems to confirm some features and problems which were referred to in the first part of this paper on the national Italian situation.

The first point we want to underline is that the supply side pioneering actors in the organic meat market in Veneto appear rather confident, if not optimistic. In fact, with reference to the last decade, we have found a meaningful dynamism in most of the firms on concerns of the breeding size and the organic meat processed quantities (Table 4). This offers hope for dealing with organic meat in the future. Most of these pioneers have declared that they have good or sufficient perspectives and particularly trust on soundness of the organic option, which some see as the more promising method for tomorrow's agriculture. Only a poultry breeder envisages limited chances for continuing organic animal production, due to the difficulties linked to this option, however, it should be noted he or she is not fully persuaded on this choice of organic.

The organic meat market niche is now smaller than the organic dairy products market and shows greater difficulties. This is typical of an unorganized supply that is still in the beginning stage. From the survey's results, it seems to emerge that in exploring this new market there is a group of operators which is heterogeneous in size, paying attention to the market (especially in regards to the selling solutions: live weight, slaughtered animal or meat cuts), costs of the new processes and, hence, economic results.

At the chain level, we found a certain contractual weakness of the breeding farms in comparison to the processing enterprises. The traditional wholesale channel shows either logistic problems or a scarce product positive valuation, which are linked also to an insufficient supply concentration at the production level. For such reason, not only the processing firms but also some breeding farms have begun to diversify the sale channels.

Direct selling from organic breeding farms, although rewarding, seems to be limited by logistic problems and is hardly affordable in the short run, given the high cost of the required investments. Moreover it will be very difficult to go beyond the current demand level (local and by aficionados). A more promising channel for the processing industry seems to be the way of the small specialized retail shops, as long as it will succeed in containing prices and it supports sales through a proper promotional strategy. It still remains an open question if LSR could be an outlet channel for organic meat produced in Veneto. By now it must be recognized that in the first Italian region for meat production, the internal supply of organic meat is in fact fairly low and the same processing operators, which supply the LSR, tend to supply themselves from outside the region, mainly abroad. A more practicable method appears to be selling meat to school caterers or restaurants and establishing stable agreements between these operators and the producers/processors of organic meat.

Among the surveyed enterprises, some of them (processing firms and some breeding farms) appear more engaged in marketing, aiming at strengthening the image which is associated to their brand among the consumers and at the development of other marketing tools. They are trying to adapt their supply, not only in quantities, but also in aspects which are related to meat quality, logistic and distribution. Others, in particular some breeding farms, seem to pay more attention to obtaining satisfactory economic results by controlling production costs.

By either considering what has emerged form the survey or relying on the opinion of qualified experts, it appears that some regulation aspects make the conversion to organic techniques very hard for almost all the operators in the conventional Venetian breeding farms. A first difficulty emerges from the ban on raising the same species following both the organic and conventional method on the same farm. This ban was written to avoid management mistakes inside the farm. In effect, many operators who would want to

Table 4. Business size variation and perspectives (entrepreneurs' answers)

		Firm type				
		Cattle breeding	Poultry breeding	Sheep breeding	Pig breeding	Meat processing
Changes in the business size in the last ten years		2 Yes	Yes/No	Yes	Yes	2 Yes
Perspectives for keeping on organic meat business:	Limited	–	Yes/No	–	–	–
	Sufficient	Yes/No	Yes/No	–	–	–
	Good	Yes/No	–	Yes	Yes	2 Yes

begin conversion gradually, face this radical choice, and they are not ready to involve the whole breeding herd, especially if it is large. Organic cattle breeding also implies organic cultivation of the farm's arable lands. There must be a connection between animals and cultivated surface since at least 35% of the ration must be produced by the same farm or by other local farms to which the breeder is linked for manure disposal. However, a single crop system based on corn is very common in the Venetian conventional livestock farms, since it provides the highest number of feed units per hectare at the lowest cost. By the contrary, organic agriculture requires the rotation of corn with other species, including some pulses. As an example, if we considered the rotation corn-barley-soybean, which is useful for weed control and for fertility conservation, we cannot forget that it gives a significant lower number of feed units. Similarly we should not forget that weed control in arable land could only partially be attained by mechanical interventions, since we cannot use chemicals on organic agriculture. All this determines a 20–30% decrease in the feedstuff production, which can be translated in a 40–50% increase in production costs for organic cattle breeding. This increase can reach 200% in the organic poultry breeding, given that it has cycles of production which are at least double (sometime also triple) in length when compared to conventional breeding.

Generally, we can assert that more intensive conventional breeding will have more significant cost differences between conventional and organic breeding. This is confirmed by the breeding farms that we have surveyed.

In fact, the sheep farm in the survey originates from a reality where it was already very close to what is required for organic livestock breeding. It was classifiable as an extensive breeding farm that used marginal areas (banks of rivers, high-water beds, mountain meadows), which does not need important off-farm inputs (concentrates, integrators, drugs) and technologically advanced structures. Therefore, production costs are substantially analogous in the two breeding systems on this farm because the only additional cost regards organic certification, which is widely compensated by the financial aid granted by the regional RDP.

On the other side of the extensive-intensive spectrum, we estimate a 40–50% production cost increase for the two organic cattle breeding farms, with the highest value in the larger one due to its livestock belonging to particularly expensive breeds. The pig breeding farm also shows similar cost increments.

Finally, in the two organic poultry breeding farms, cost variations are still more remarkable, in particular for broilers whose production cycles range from 100 to 120 days, which is much longer than in the conventional breeding. For this reason there is greater concentrate consumption and higher labor need, given that the very simplified buildings for organic production (although providing medium depreciation costs) force a breeder to manually execute a series of operations, unlike the fully mechanized sheds of the conventional breeders.

Another important aspect for cattle breeding regards stocking. Its maximum level for organic production is conditioned by the 170 kg of nitrogen per hectare constraint, which implies from 5 to 3.3 heads of cattle per hectare in relation to their weight. These stockings rates are much lower than those we can find in most of the Venetian flatland breeding farms, which have small surfaces so the number of head rises higher than what organic regulations allow. Therefore, if the breeder wants to convert to organic breeding, he or she would be forced to reduce the number of head with the same land or to increase the available land, a difficult act to carry out in the current Veneto land market.

The supply of organically reared young animals is another crucial point. At present, it is practically impossible to find calves that are certified as produced by an organic breeding farm. Moreover, if the organic breeder needs to rear animals in a open stall system, based on structures with full paving (namely without grated floor), and must offer a pasturing possibility, the current situation of Venetian livestock breeding activity makes the conversion process even more difficult to undertake.

In conclusion, rearing animals in compliance with the principles of organic animal breeding implies higher costs for a decrease in crop yields, higher per head fixed costs and higher feeding costs if one needs to buy on the organic feed market. A reasonable acknowledgment of the entrepreneurial choice carried out by organic meat producers should yield at least a 40–50% premium price for beef meat and a 150–200% premium for poultry meat. But this would result in a price that only few consumers would be willing to pay since most consumers are only willing to pay 25–30% more for organic meat.

It is clear that the high cost variations prevent the conversion to organic meat production. Even for operators who would be well disposed to it are hesitating in front of a market, which, at present, is not able to guarantee a sufficient boost in the premium price. In this scenario, we think that a species-related subsidy calibrated on the per head cost of production increase due to conversion to organic breeding, should be integrated into the general per hectare aid provided by the Veneto Region RDP for organic crop cultivation. This would provide a support tool that is more effective and more targeted towards organic breeding.

Previous comments noted that Venetian organic meat processors have to import meat from other EU countries because the production costs are lower than in Italy. In fact, in these production areas in other countries, although they are difficult and marginal for agriculture, organic breeding costs are only about 20% higher than those for conventional breeding. They are unquestionably lower than those we can find in conventional breeding farms of Veneto flatland. Hence, these operators are able to acquire organic certified meat with a 20–25% premium price over the conventional one.

The organic meat market at the regional level still needs to be defined. Currently organic meat is available in four specialized butcheries (one in Verona, two in Padua, one in Rovigo), in about thirty organic specialized shops and is beginning to come out in some conventional butcheries and in LSR outlets. Although, interest seems to increase, given the expectations of consumers who are looking at these products as remedy against recent food scandals fears, organic meat consumption ranges between 1.5 and 2% of total meat retail value. Gross sales can be estimated to reach 10 million EUR, which is a rather low value typical of an industry that is still in the beginning stage and where price is the key factor that limits consumption. This is a problem that affects the whole organic food industry. On the other hand, price reductions have taken place during the last two years, following the increasing LSR interest. Nevertheless, difficulties are surely greater for organic meat, given that from one side it is more difficult to reduce costs and, from the other, meat has a broad share in the expenditure for conventional food.

Now, after three years, the firms which first began to operate in the organic meat industry are beginning to see the first results, while the others are investing and consider themselves in the launching phase. So, even with the BSE crisis and the new EU industry regulations, many things still remain to be built in this industry, either at the production level (breeding technologies, *ad hoc* structures, chain integrations, identification of the most suitable places) or at the market level (promotional campaigns and consumer's information).

At the end of this first analysis of Venetian pioneering firms in the organic meat industry, we want to stress that further research is needed, especially about technical and economic feasibility of organic breeding which can be suitable for the Venetian agriculture conditions. Special attention is needed for some zones (mountain, park areas) that may find organic breeding the proper balance between production and environment safeguard.

In spite of current difficulties, the survey has shown interesting elements that could considerably modify the industry's outlook in the future. They regard the possibilities that feedlot based cattle breeding, which is the prevalent meat production system in Veneto, can find a solution to problems such as stocking, organic calf supply, feed cost reduction and other market-related problems. The first research undertaken in this field shows that breeding costs can be controlled to a range of 20% to 25% increase over conventional meat. Hence, organic meat should have a reliable market perspective of achieving a 30–50% premium price and a strong loyalty from the Italian consumer.

4. REFERENCES

Ansaloni, F., and Bellavia, R., 2001, Commercializzazione di bovini da carne biologici, *L'informatore Agrario*, **3**: 35–38.
Baron, F., 2000, Il metodo biologico entra nella stalla, *Largo consumo*, **5**: 49–50.
Cozzi, G., Preciso, S.F., Gottardo, F., and Andrighetto, I., 2001, L'allevamento biologico come alternativa ai sistemi intensivi di produzione della carne bovina, *L'Informatore Agrario*, **17**: 101–107.
Databank (2001), *Prodotti alimentari biologici nella distribuzione moderna—Informazioni Base di Settore*, Milano.
Didero, L., 2001a, Novità in campo, *Largo consumo*, **3**: 99–103.
Didero, L., 2001b, Biologico in tutte le mense, *Largo consumo*, **7/8**: 102–105.
Didero, L., and Troiani, C., 2001, Fuori dalla nicchia, *Largo consumo*, **7/8**: 62–66.
Di Marco, A., 2001, Zootecnia bio a un bivio, *Bioagricultura*, **70**; http://www.aiab.it/nuovosito/informazione/bioagricultura/articolo.php/68 .
Lunati, F., 2001a, Valutazioni sull'andamento dell'agricoltura biologica in Italia, Paper presented at the Meeting *L'evoluzione del Biologico in Italia*, Bologna, September 13, 2001.
Lunati, F., (ed.), 2001b, *Il biologico in cifre, Rapporti Biobank*, Distilleria EcoEditoria, Forlì.
Marengo, G., Bassignana, E., Corsi A., and Didero, L., 2000, *Le prospettive del mercato dei prodotti zootecnici da agricoltura biologica*, Regione Piemonte—Assessorato Agricoltura, Torino.
Menghi, A., De Roest, K., and Torelli, F., 2001, Fettine biologiche ad alto gradimento, *Largo consumo*, **2**: 27–31.
Rossi, P., and Gastaldo, A., 2001, Costo delle strutture per l'allevamento biologico dei suini, *L'Informatore Agrario*, **14**: 39–44.

PROFITABILITY OF ORGANIC CROPPING SYSTEMS IN SOUTHWESTERN MINNESOTA[*]

Paul R. Mahoney, Kent D. Olson, Paul M. Porter, David R. Huggins, Catherine A. Perillo, and R. Kent Crookston[**]

SUMMARY

In spite of concerns, Minnesota's dominant cropping system is the corn-soybean rotation using synthetic pesticides and chemically processed fertilizers. Using experimental data from 1990–1999, this study compared the profitability of organic versus conventional strategies. Net return (NR) was calculated from actual yields, operations, inputs, prices, and organic premiums. Yields and costs were lower for the 4-year organic strategy. With premiums, the 4-year organic strategy had NRs significantly higher than conventional strategies; without premiums, the NRs were statistically equal ($p = 0.05$). Thus, the 4-year organic strategy was not less profitable nor its NR more variable than the conventional strategies in this study.

1. INTRODUCTION

By most measures, crop production systems are more efficient and productive today than at any time in the past. This is due in large part to improved crop varieties, improved

---

[*] This work was supported by the University of Minnesota and a cooperative agreement with the Economic Research Service, U.S. Department of Agriculture. Originally published as Mahoney, Olson, Porter, Huggins, and Crookston. Profitability of organic cropping systems in southwestern Minnesota. Renewable Agriculture and Food Systems, 19(1):35–46, 2004. Permission granted for reproduction by CABI Publishing Wallingford Oxon,UK.

[**] Paul R. Mahoney works with AgCountry FCS in Morris, MN, and a former Regional Extension Educator and former graduate student at the University of Minnesota; Kent D. Olson, and Paul M. Porter are, respectively, a professor at the Department of Applied Economics, and an associate professor in Agronomy and Plant Genetics, University of Minnesota; David R. Huggins is a soil scientist with the USDA-ARS, Pullman, WA; Catherine A. Perillo is an instructor, Department of Crops and Soil Sciences, Washington State University; R. Kent Crookston is dean of the College of Biology and Agriculture, Brigham Young University, Provo, Utah.

farm equipment, better management skills, synthetic pesticides, and chemically processed fertilizers. While the use of synthetic pesticides and chemically processed fertilizers has contributed significantly to gains in productivity, they have also raised concerns from the general public about food safety and adverse environmental quality effects. In addition, the current price and income situation have increased farmers' interest in organic production methods.

However, the vast majority of farmers in southwest Minnesota (and the Midwest) continue to produce crops with a traditional corn-soybean rotation using production practices which involve, at some level, chemical use and commercial fertilizers. Farmers' reasons for not changing from traditional cropping systems are as diverse as the farmers themselves. Some of the reasons include the uncertainty of the profitability of organic systems, increased labor that may be required by an organic system, lower yields with other systems, the cost in money and time to learn other systems, and the difficulty of finding markets for organic products.

Previous studies have analyzed the profitability, sustainability, and yields of organic farming practices, and have shown generally that organic systems can be as or more profitable than conventional systems (Roberts and Swinton, 1996; Welsh, 1999). Although price premiums paid for organic products increase profitability and are highly sought after, Welsh found in his research review that they were "not always necessary for organic systems to be competitive with or outperform conventional systems" (Welsh, 1999, p.40). However, these studies also found that organic systems were not necessarily without potential pitfalls and not necessarily for every farmer and farm site.

In response to the concerns about the impact of conventional farming practices, the University of Minnesota started the Variable Input Crop Management Systems (VICMS) study in 1989 to estimate and compare the agronomic and economic impacts of organic production methods in southwest Minnesota. To help decrease some of the uncertainty regarding profitability of organic production methods, this current study used data from the VICMS study to compare the profitability and riskiness of three different management strategies, two cropping sequences, and three organic price scenarios. The value of this study comes from the use of a long-term study (10 years) in which each crop in each management strategy and each sequence is present and replicated three times in each year. Due to the continued dominance of conventional production systems, the current study's primary objective was to test the hypothesis that conventional systems were more profitable and/or had less risk than organic systems.

2. BACKGROUND

Sales of organically grown products have steadily increased over the past two decades (Duram, 1998). Even though the numbers are small relative to the entire food market, the number of organic products on grocers' shelves and the amount of grocery shelf space being devoted to organically produced products has increased. As consumers become more concerned about food and environmental safety, they also become more concerned about the practices and inputs being used to produce the food they consume. Since some consumers are willing to pay a premium for products produced by organic methods, there appears to be a separate demand for organic products in the marketplace. In addition, as Krissoff (1998, p.1131–2) notes, "increasing farm, agribusiness, and food marketing investments in alternatively produced products suggest a responsiveness to the growing interest in the organic foods".

In general, organic systems do not use synthetic pesticides and chemically processed fertilizers and have longer crop sequences with a greater variety of crops being grown. However, the specific methods, certification of organic methods, and labeling of organic products were not uniform across the U.S. until the National Organic Program (NOP) rules were finalized in December 2000, and fully implemented in October 2002. Under the NOP rules, all organically labeled products must be produced and processed using a set of specific standards to ensure consistent practices nationwide. Organic producers must operate under an organic system plan approved by an accredited certification agency or agent. Under these standards, if a product is to be sold and labeled as organically produced, all materials used in production must be used in accordance with the National List of Allowed Synthetic and Prohibited Non-Synthetic Substances. Specifically, the NOP crop production standards state that: "Land will have no prohibited substances applied to it for at least 3 years before the harvest of an organic crop. The use of genetic engineering (included in excluded methods), ionizing radiation and sewage sludge is prohibited. Soil fertility and crop nutrients will be managed through tillage and cultivation practices, crop rotations, and cover crops, supplemented with animal and crop waste materials and allowed synthetic materials. Preference will be given to the use of organic seeds and other planting stock, but a farmer may use non-organic seeds and planting stock under specified conditions. Crop pests, weeds, and diseases will be controlled primarily through management practices including physical, mechanical, and biological controls. When these practices are not sufficient, a biological, botanical, or synthetic substance approved for use on the National List may be used" (United States Department of Agriculture, 2001).

3. STUDY LOCATION AND DESIGN

The VICMS study was situated at the University of Minnesota's Southwest Research and Outreach Center near Lamberton, Minnesota, about 240 kilometers southwest of Minneapolis-St. Paul. In this area of Minnesota, crop production began in the 1870s with wheat grown almost exclusively. From the 1900s until the 1960s, corn, small grains, and pasture predominated. Since the 1960s, this region had been farmed almost exclusively with corn and soybean. Recently, corn and soybean were grown on more than 90% of the cropped land in Southwest Minnesota (Minnesota Agricultural Statistics Service, 2001).

The data for this study were from part of the study called VICMS II that had been cropped according to University recommendations since 1959 resulting in high soil fertility levels and low weed populations. Since the common soil condition in this part of Minnesota was high fertility and low weed pressure, the VICMS II data was important for producers interested in the transition from conventional practices to organic practices. Data from 1990–1999 were used in this study. Although they may differ from actual farm yields, the yields obtained in these research trials were considered the best comparison of different management strategies and cropping sequences due to being obtained on the same location with all crops present in every year.

The three management strategies analyzed in this study used a low level of purchased inputs, a high level of purchased inputs, and organic inputs. Both high and low levels were included since the set of practices thought of as "conventional" was similar to the high level of inputs at the beginning of the study but had moved closer to the low level of inputs by 1999. A general description of each strategy is given below with a general description of the production practices followed in each strategy listed in Table 1.

- *Low-Purchased Inputs (LI).* Chemical applications were minimized by banding of fertilizers, banding of post-emergent herbicides (if needed), utilization of mechanical weed control, use of insecticides only if prescribed, and similar practices. A realistic yield goal was used to determine fertilizer rates. This realistic yield goal was based on soil type, water availability, growing season length, and actual recorded yields in the past, not an optimistic view of the soil potential.
- *High-Purchased Inputs (HI).* Chemical applications were not necessarily minimized. Broadcast (no banding) fertilizers and insecticides were used according to University recommendations. Pre-emergent herbicides were often used. Other practices were selected on the basis of what was considered the best conventional practices for this region. An optimistic yield goal was used to determine fertilizer rates. However, even this optimistic yield goal was "realistic" in that it too was based on actual recorded yields in the past, not an optimistic view of the soil potential.
- *Organic Inputs (OI).* No synthetic chemical applications were used. Organic sources of nutrients, such as manure, and mechanical weed control were utilized. The OI strategy incorporated the best organic practices for the region based on practices approved by a designated certification organization recognized by the Minnesota Department of Agriculture. The OI crops were not but could have been certified under NOP rules. Potential organic premiums were not applied until certification was possible under organic certification standards (i.e., the third crop) in the 4-year sequence.

These three management strategies were carried out in two cropping sequences: the popular two-year sequence (corn-soybean) and a four-year sequence (corn-soybean-oat/alfalfa-alfalfa). Rotation restrictions did not allow the certification of a 2-year corn-soybean sequence as organically produced so, even though an organic management strategy was followed, crops grown under the 2-year sequence did not receive an organic premium in any year. Every crop in each cropping sequence was grown every year under each management strategy, so all treatments were present each year. Each combination of strategy and sequence was replicated three times in each year.

4. DATA COLLECTION AND ANALYSIS METHODS

Detailed records have been maintained on field operations, labor used, rainfall, plant growth, weed counts (broadleaf and grasses separately), earth worm species and counts, mycorrhiza in the soil and plants, and crop yield. Soil P and K fertility levels were determined in the fall, and soil nitrate levels were determined in 0.304 meter increments to 1.52 meters following alfalfa and soybeans.

The net return (NR) was calculated as gross income from the crops produced minus the direct production costs for each management strategy, cropping sequence, and year. NR was expressed as the return per one hectare averaged over either 2 or 4 crops. The

Table 1. General description of the three management strategies: Low-Purchased Inputs (LI), High-Purchased Inputs (HI), and Organic Inputs (OI)[a] for four crops

Practice	LI	HI	OI
Corn			
Prior fall tillage[b]	None	Chisel	Chisel
Spring tillage	Field cul. (2x)	Field cul. (2x)	Field cul. (2x)
Rotary hoeing	1–3x (as needed)	None	1–3x (as needed)[c]
Row cultivation	2–3x	1–3x	2–3x
Tillage after harvest	MB*	MB	MB
Herbicides	Pre-e post.	Pre-e post.	Organic
Fertilizer application	Banded	Broadcast	Organic
Soybean			
Prior fall tillage	Soil saver	MB	MB
Spring tillage	Field cul./disk	Field cul./disk	Field cul./disk
Rotary hoeing	1–2x	None	1–2x
Row cultivation	2–3x	1–2x	2–3x
Tillage after harvest	None	Chisel	Chisel
Herbicides	Pre-e post.	Pre-e post.	None
Fertilizer application	Banded	Broadcast	Organic
Oat			
Prior fall tillage	None	Chisel	Chisel
Spring tillage	Field cul. (1x)	Field cul. (1x)	Field cul. (1x)
Rotary hoeing	None	None	None
Row cultivation	None	None	None
Tillage after harvest	None	None	None
Herbicdes	None	None	None
Fertilizer application	Broadcast	Broadcast	Organic
Alfalfa			
Prior fall tillage	None	None	None
Spring tillage	None	None	None
Rotary hoeing	None	None	None
Row cultivation	None	None	None
Tillage after harvest	MB	MB	MB
Herbicdes	None	None	None
Fertilizer application	Broadcast	Broadcast	Organic

[a] Specific operations used each year may have been different.
[b] In the 4-year rotation the previous fall tillage was moldboard plowing (MB) of the alfalfa residue.
[c] For the OI strategy, the fertilization was with fall-applied, composted beef manure for the 4-year rotation, and spring-applied swine manure for the 2-year rotation.
* MB = moldboard plowing.

value of corn and soybean was included in NR for the 2-year sequence, and the value of corn, soybean, oat, oat straw, and alfalfa was included for the 4-year sequence. The gross income of each crop was calculated by multiplying the actual crop yield by the typical harvest cash price plus a potential organic premium (if applicable). The yield was the average of three replications. Typical harvest cash prices were those reported annually by the Southwestern Minnesota Farm Business Management Association (SWMFBMA, Table 2). Harvest prices were used (versus annual average prices) to account for the value of production only and not for marketing skill. These SWMFBMA prices were used since they were considered to reflect more accurately the prices faced by farmers

Table 2. Typical harvest cash prices from the Southwestern Minnesota Farm Business Management Association and all Minnesota (USD/Mg)

Crop/Year	1990	1991	1992	1993	1994	1995	1996	1997	1998	1999
Corn										
SW Ass'n	78.75	82.69	70.87	88.59	70.87	108.28	94.50	94.50	68.91	68.91
Minnesota	81.11	84.66	75.21	83.47	75.60	100.41	102.77	92.53	68.12	59.85
Soybean										
SW Ass'n	211.31	92.94	192.94	220.50	183.75	211.31	257.25	238.87	189.26	189.26
Minnesota	212.78	199.92	193.67	216.09	198.45	212.41	264.60	234.46	182.65	158.02
Oats										
SW Ass'n	86.13	68.91	68.91	86.13	75.80	103.36	137.81	137.81	82.69	82.69
Minnesota	72.35	68.22	83.38	88.89	74.42	94.40	135.06	103.36	67.53	60.64
Alfalfa										
SW Ass'n	66.15	55.13	60.64	77.18	77.18	77.18	88.20	104.74	71.66	71.66
Minnesota	104.74	73.32	88.48	105.01	82.41	80.21	92.61	115.49	88.48	72.49
Oat Straw (USD/bale)										
SW Ass'n	1.00	1.00	1.00	1.50	1.50	1.75	2.00	2.00	1.00	1.50

Sources: Adapted from annual association reports (e.g., Olson et al., 1992) and annual *Minnesota Agricultural Statistics* (e.g., Minnesota Agricultural Statistics Service, 2001).

close to the VICMS study site compared to using an overall Minnesota average price. Except for alfalfa and the oat price in later years, the SWMFBMA prices followed the annual levels and patterns of average harvest-time prices in Minnesota. Direct production costs were determined by the actual operations, input levels used, and local input prices. The costs for land, management, and indirect costs (farm insurance, marketing, and so on) were not subtracted from the gross value of the crops because they would not vary between strategies and sequences. Thus, NR was the net return to land, management, indirect labor, and other indirect costs for each management strategy and cropping sequence.

Production costs were estimated for each year using the actual cultural operations, equipment and inputs used, as listed in the research field records for each strategy and sequence. The cost of each operation was calculated using the University of Minnesota Extension Service's annual estimates of machinery costs, which included fuel, maintenance, repairs, operator labor and overhead costs of the machinery (e.g., Fuller et al., 1992). Each strategy and sequence was charged the same cost per pass for the same operation (cultivating, for example), but the total cost per hectare varied when the number of passes varied (2 cultivations versus 1 cultivation, for example). Labor costs for machinery operations were included in extension's estimates, but since no records were kept to show differences in time required for pest scouting, managing, marketing and other indirect labor use, these labor costs are not included in the calculation of NR. Manure application costs were included in the organic systems, but no manure purchase cost was included. Local market prices were used for inputs except for seed and herbicide. Seed costs were taken from annual SWMFBMA records (e.g., Olson et al., 1992). Herbicide prices were taken from University of Minnesota Extension Service's annual weed control report (e.g., Durgan et al., 1992). Total crop production costs were the sum of tillage, planting, fertilizer, pest control, harvesting, and hauling costs.

Table 3. Average organic price premium ratios based on organic price quotes and U.S. cash prices

Year	Corn	Soybeans	Oats
1995	1.35	2.14	1.35
1996	1.43	1.85	1.59
1997	1.73	2.41	1.73
1998	1.88	3.02	1.83
Average ratio	1.60	2.36	1.63
Average premium	60%	136%	63%
Half of the average premium	30%	68%	31.5%

Note: Due to insufficient data on organic prices for alfalfa and oat straw, there were no organic premiums estimated for these two crop products.
Source: Dobbs and Pourier, 1999.

Due to insufficient annual information on organic premiums in Minnesota (and the U.S.) for 1990 through 1999, potential organic premiums were estimated for corn, soybean, and oat using information compiled by Dobbs and Pourier for 1995 through 1998 (1999, Table 3). Over these four years, the average price for certified organically produced corn was 160% of the average U.S. cash price for non-organic corn; that is, the average premium for organic corn was 60% over the average non-organic corn. The average price for certified organically produced soybean was 236% of the average U.S. cash price for non-organic soybean; that is, the average premium for organic soybean was 136% over non-organic soybean. The average price for certified organically produced oat was 163% of the average U.S. cash price for non-organic oat; that is, the average premium for organic oat was 63% over the average non-organic oat. In the absence of reliable annual data on organic prices, these average organic premiums were applied to the typical harvest cash prices for corn, soybean, and oat in the 4-year, OI strategy starting in 1992, the third year of production and the first that could be certified as organic. Production in 1990 and 1991 did not receive an organic premium because the land had not been in a certified organic production system for the required 36 months.

This pattern of not applying organic premiums in the first two years but then applying them to eligible production in the third and following years simulated the transition a conventional farmer would have to go through to sell organically produced crops as certified organic. Due to lack of data, no price premium was considered for either organic alfalfa or organic oat straw. Because it would not meet NOP rules for certification, crops grown under the 2-year sequence did not receive organic premiums in any year even though all other organic practices were followed. Using the same average organic premium for all years reduced the potential variability of NRs; however, the alternative of using annual premiums based on very little data seemed much less desirable.

Potential organic premiums vary from year to year and are also dependent on each individual producer's marketing strategies and abilities. To reflect this variability, three organic price scenarios were evaluated as well as the sensitivity to the oat straw price. In the first scenario (as just discussed in the previous paragraph), corn, soybean, and oat grown under the 4-year, OI strategy received the full historical average organic premium starting in the third year, i.e., 1992. The second price scenario was more conservative and assumed that eligible crops received only half of the historical average organic premiums starting in 1992 (or that only half of the eligible production received the historical average). The third price scenario contained no organic premiums even for certified organic production; every crop in every management strategy and sequence received the typical harvest cash prices reported earlier. This last price scenario was useful to evaluate the situation that may occur if a farmer could not obtain the organic premium or if the supply of organic production increased faster than the demand and put downward pressure on the organic premium. In all three price scenarios, non-organic crop production from the 2-year OI strategy and both sequences of the HI and LI strategies received only the typical harvest cash prices in every year. To evaluate whether the straw value would affect a farmer's choice of production method, the oat straw price was set to zero and NRs recalculated in a subsequent evaluation for all three scenarios.

Risk, that is, the variation in NR, was analyzed using stochastic dominance. Cumulative distribution functions (CDFs) of NRs were calculated based on the yields, market prices, input costs, potential organic premiums, correlations between crop yields, and correlations between crop yield and market price. The CDFs were calculated using a program called Crystal Ball © (CB) which is an add-in program that functions within Microsoft's Excel ©. Crystal Ball was used to develop a probability distribution of net returns based on the averages and distributions of yields and market prices, average input costs, average potential organic premiums, correlations between crop yields, and correlations between crop yields and market prices. The distribution assigned to each individual crop yield within each strategy and sequence was based on actual recorded yield data with the Kolmogrov-Smirnov (KS) test used to determine the best fitting distribution. Eleven distributions (Normal, Lognormal, Weibull, Triangular, Uniform, Beta, Exponential, Gamma, Logistic, Pareto, and Extreme Value) were considered. The top three best-fitting distributions (based on the KS Test) were compared visually to the distribution of the actual yields as a visual check of their goodness-of-fit. For crop yields, 30 yield observations (3 reps/crop for 10 years) from each crop were used in fitting the distributions; if available, 33 observations from 11 years were used. Using the same methods as for yields, crop price distributions were estimated using data from 1990 through 1999.

The total input cost of each crop in the risk analysis of the study was assumed to be constant based on the actual historical 10-year average of input costs. Average projected costs were used since historical input prices, actual yields, and field operations were used in the calculation of the yearly input costs. By using the 10-year average of input costs, it more accurately reflects the relationships and actual decisions made historically. Input costs in this part of the analysis were held constant because this is most likely the way individual farmers would represent their own costs in a similar forecasting or budgeting scenario. Potential organic premiums (i.e., ratios of organic prices to U.S. cash prices) were also considered constants due to the lack of adequate data to estimate a sound distribution based on the historical data.

Correlations between crop yield and price were calculated using actual crop yields and their respective crop prices (Table 4). Correlations between crops were also calculated using the actual recorded crop yields from the VICMS II data correlating the corn yield to other crops in the sequence (i.e., soybeans in the 2-year sequence and soybeans, oats, and alfalfa in the 4-year sequence; Table 5).

Using the assigned distributions of crop yields and crop market prices, CB calculated 500 different possible random draw combinations of crop yields and prices. Using these 500 possible outcomes of yield and price, in addition to input costs, and potential organic premiums, 500 possible outcomes of NR for each input strategy and cropping sequence

Table 4. Price/Yield correlations for each crop in each cropping sequence and management strategy from 1990 through 1999 in the VICMS II study

Strategy	Corn 2-year	Soybean 2-year	Corn 4-year	Soybean 4-year	Oat 4-year	Alfalfa 4-year	Oat Straw 4-year
LI	−0.847	0.059	−0.700	0.049	−0.252	−0.020	−0.525
HI	−0.683	−0.127	−0.734	−0.024	−0.341	−0.004	−0.525
OI	−0.197	−0.226	−0.557	−0.260	−0.314	0.091	−0.537

Source: Estimated from actual VICMS II experiment yield data and the prices listed in Table 2.

Table 5. Yield correlations for each crop in each cropping sequence and management strategy from 1990 through 1999 in the VICMS II study

Strategy	Corn/SB 2-year	Corn/SB 4-year	Corn/Oat 4-year	Corn/Alfalfa 4-year	SB/Oat 4-year	SB/Alfalfa 4-year	Oat/Alfalfa 4-year	Oat/Oat Straw 4-year
LI	0.643	0.558	0.369	−0.127	0.401	−0.479	−0.672	1.000
HI	0.713	0.456	0.353	−0.163	−0.080	−0.259	−0.570	1.000
OI	0.338	0.038	0.069	−0.180	0.328	−0.166	−0.667	1.000

Source: Estimated from actual VICMS II experiment yield data.

were also calculated. The 500 estimated NRs for each cropping sequence and input strategy are then used to develop their respective CDFs.

First-degree stochastic dominance (FSD) and second-degree stochastic dominance (SSD) were used to determine farmers' potential risk preferences between input strategies and cropping sequences (Hardaker et al., 1997).

5. RESULTS

Annual corn and soybean yields in the 4-year sequences under the LI, HI, and OI strategies tend to be equal to or better than the corn and soybean yields in the 2-year HI and LI strategies (Figures 1 and 2). However, the average yields were not substantially different except for the 2-year OI strategy, which had lower yields (Table 6). Although they are not shown due to space constraints, annual oat and alfalfa yields in all strategies followed very similar patterns. Only in 4-year OI corn could a potential transition period be seen due to its lower yields relative to other strategies (except 2-year OI), however, this period was uncertain due to all strategies and sequences having lower yields in the fourth, extremely wet year. For other crops, the 4-year OI yields follow the annual patterns similar to other strategies (Figures 3 and 4). Annual VICMS yields for all crops follow a similar pattern compared to the average yields reported for this part of Minnesota by the USDA. In most years, VICMS yields were higher than the USDA averages with two exceptions. First, the organic yields (especially the 2-year rotation) were lower than average yields in this area. Second, the VICMS alfalfa yield was

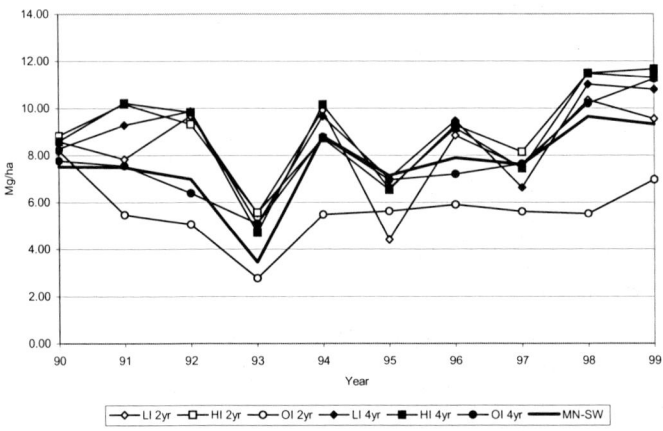

Figure 1. Corn yield by cropping sequence and management strategy, 1990–1999

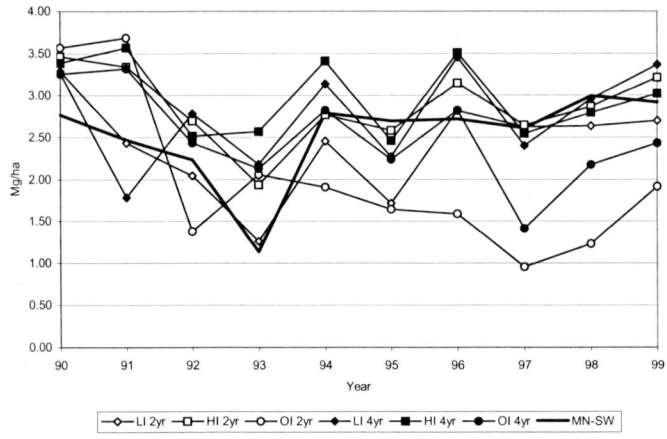

Figure 2. Soybean yield by cropping sequence and management strategy, 1990–1999

Table 6. Average crop yields for each crop in each cropping sequence and management strategy from 1990 through 1999 in the VICMS II study (Mg/ha)

Strategy	Corn 2-year	Soybean 2-year	Corn 4-year	Soybean 4-year	Oats 4-year	Alfalfa 4-year
LI	8.22	2.42	8.72	2.75	2.22	11.2
	(1.95)*	(0.58)	(1.92)	(0.57)	(1.00)	(2.01)
HI	8.90	2.89	8.97	2.96	2.29	11.4
	(1.88)	(0.45)	(2.21)	(0.46)	(1.08)	(3.36)
OI	5.64	2.02	7.90	2.49	2.29	11.6
	(1.38)	(0.93)	(1.79)	(0.57)	(1.14)	(2.91)

* Standard deviations are in parentheses.

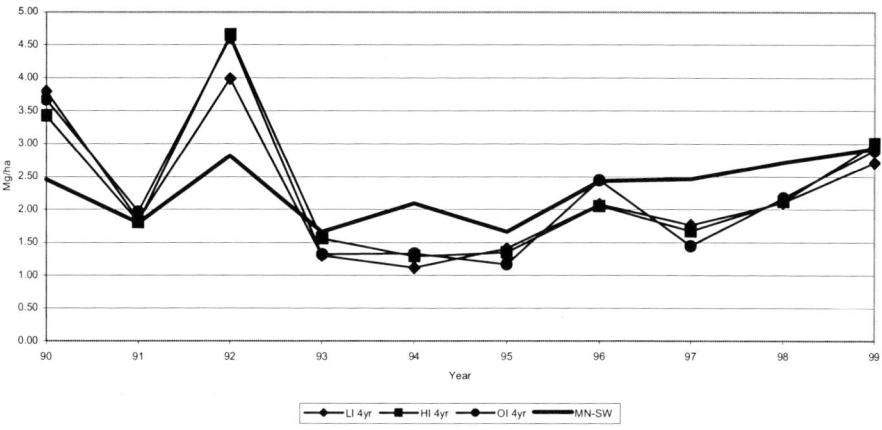

Figure 3. Oat yield by cropping sequence and management strategy, 1990–1999

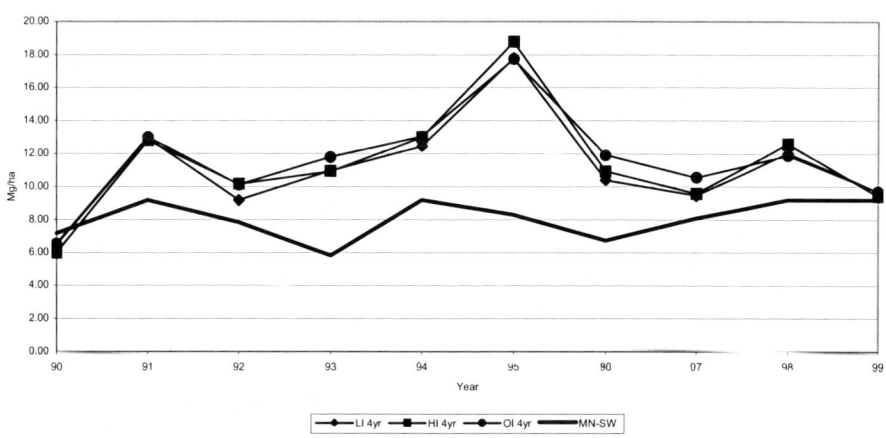

Figure 4. Alfalfa yield by cropping sequence and management strategy, 1990–1999

obviously higher in most years than the USDA average. The higher alfalfa yields may be due to the VICMS study site being on better soils than most alfalfa fields in southwest Minnesota and/or due to more management time spent on alfalfa within VICMS than an average farmer would spend. Porter *et al.* (2002) provided a more detailed analysis of the agronomic results in the VICMS study.

Relative to the yields under conventional management, VICMS organic yields were similar to those obtained in other studies in the Midwest. In the VICMS II study, the 4-year OI corn yield was 89% of the 2-year HI corn yield and the 4-year OI soybean yield was 86% of the 2-year HI soybean yield. In a study at Iowa State, the 10-year average organic corn yield was 81% of the conventional corn yield; they did not include soybean in their organic rotation (Welsh, 1999, p. 24). At Nebraska, the 8-year average organic corn yield was 85% of the conventional corn yield and the 8-year average organic soybean yield was 84% of the conventional soybean yield (Welsh, p. 32). At South Dakota State, the 7-year average organic corn yield was 84% of the conventional corn yield and the 7-year average organic soybean yield was 99% of the conventional soybean yield (Welsh, p. 34).

In both the 2- and 4-year sequences, the HI strategy had the highest average direct production costs compared to the LI and OI strategies (Table 7). The OI strategy had the lowest costs. Direct production costs for corn in the 2-year sequence average $116 higher per hectare under the HI strategy than under the OI strategy. In the 4-year sequence, the production costs for corn averaged $89 higher under the HI strategy than under the OI strategy. The variation between years was also higher for all crops under the HI strategy as shown by the higher standard deviations. The lower costs for the OI strategy were primarily due to no expenditures for synthetic pesticides and chemically processed fertilizers even though it did pay more for additional mechanical weed control.

When full historical average organic price premiums (starting in the third year according to NOP rules) were applied to corn, soybean, and oat grown under the 4-year OI strategy, the 10-year average NR was $667 per hectare (Table 8). This was significantly ($p = 0.05$) higher than the HI and LI strategies in both the 2- and 4-year sequences. The 4-year HI strategy had an average NR of $425 per hectare, and the 4-year LI strategy had an average NR of $427 per hectare. The 2-year HI strategy had an average NR of $378 per hectare, and the 2-year LI strategy had an average NR of $338 per hectare. Although they look different, the average NRs for the 4-year HI and LI strategies were not significantly ($p = 0.05$) different from the 2-year HI and LI strategies.

The 2-year OI strategy did not meet NOP rules for certification and, thus, could not receive any organic premiums in any year. Its average NR was $227 per hectare, which was significantly ($p = 0.05$) lower than any other strategy and sequence except for the 2-year LI strategy.

If only half the historical average organic premiums were received (or half of the production received the historical premium), the average NR for the 4-year OI strategy was $551 per hectare. This was statistically equal to the 4-year OI strategy when it received the full historical organic premium but significantly ($p = 0.05$) higher than all other strategies and sequences.

If the 4-year OI strategy did not receive any organic premiums, its average NR was $432 per hectare. This was not significantly ($p = 0.05$) different from the 4-year HI and LI strategies and seemingly higher but not significantly ($p = 0.05$) different from the NRs received from the 2-year HI and LI strategies.

The annual movements in the NR echo the similarities and differences noted in the averages just discussed. For clarity in the graph, only selected strategies and sequences were chosen to highlight the annual patterns (Fig. 5). The effect of starting to receive the organic premium in the third year was obvious as the annual NRs are substantially higher for the 4-year OI strategy compared to the other NRs. The higher levels of NR for the 4-year sequences can also be seen compared to the 2-year HI strategy.

PROFITABILITY OF ORGANIC CROPPING SYSTEMS

Table 7. Average crop production costs per hectare for each crop in each cropping sequence and management strategy from 1990 through 1999 in the VICMS II study

Strategy	Corn 2-year	Soybean 2-year	Corn 4-year	Soybean 4-year	Oats 4-year	Alfalfa 4-year
LI	291 (28.7)*	190 (16.3)	294 (28.7)	190 (18.0)	205 (32.1)	247 (41.7)
HI	358 (39.0)	203 (18.3)	351 (47.7)	217 (27.9)	222 (35.1)	257 (55.3)
OI	242 (16.5)	180 (24.0)	262 (13.8)	185 (14.3)	170 (16.5)	225 (35.3)

* Standard deviations are in parentheses.

Table 8. Average net returns for each crop in each cropping sequence and management strategy from 1990 through 1999 in the VICMS II study (USD/hectare)

Strategy & pricing alternative	2-year sequence		4-year sequence	
LI	339.4	$(120.8)^{a,c}$*	426.3	$(91.1)^a$
HI	378.4	$(108.4)^a$	423.9	$(80.5)^a$
OI with average OP**	***		667.4	$(188.2)^b$
OI with half the average OP	***		550.1	$(130.4)^b$
OI with no OP	227.7	$(121.5)^c$	433.0	$(84.5)^a$

* Standard deviations are in parentheses. Different letter superscripts indicate statistically significant differences between means ($p = 0.05$).
** OP = organic premiums.
*** Since the 2-year OI strategy did not meet the organic certification rules, it could not receive organic premiums.

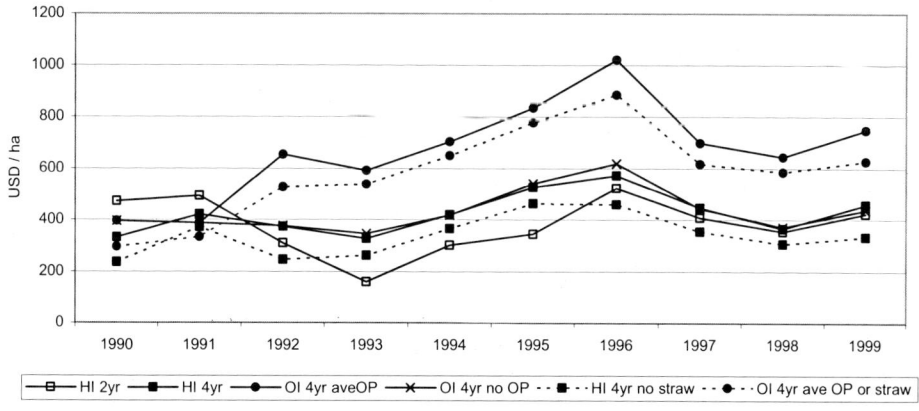

Figure 5. Annual net returns, selected strategies and sequences, 1990–1999

When the oat straw price was set equal to zero for all 4-year sequences, the general level of NR decreased, but only one change in significance occurred. The only change was for the 4-year OI strategy with half of the historical organic premium whose NR became statistically (p = 0.05) the same as the 4-year OI strategy with no premium and the HI and LI strategies in both the 2-year and 4-year sequence. Even with an oat straw price of zero, the NR for the 4-year OI strategy with the full historical premium remained significantly (p = 0.05) higher than all other strategies and sequences.

When all input strategies received conventional prices in the 2-year sequence, both the HI and LI strategies dominated the OI strategy by SSD, and the HI strategy dominated the LI strategy by FSD (Figure 6). Therefore, under the 2-year sequence, with all input strategies receiving the same market prices, the HI strategy would be preferred over the OI and LI strategies.

With all input strategies receiving the same conventional market prices (including the initial positive straw prices) under the 4-year sequence, the CDF of the OI strategy was equal to or strictly below and to the right of the CDF of the HI strategy, thus dominating the HI strategy by FSD under the 4-year sequence (Fig. 7). Under the 4-year sequence there was no FSD or SSD between the OI and LI strategies. Although it is not clearly visible in Figure 7, the LI strategy had a lower NR than the HI strategy at a probability level of 1% or less (i.e., the CDF of the LI strategy begins to the left of the CDF of the HI strategy), therefore there was neither FSD nor SSD between the LI and HI strategies. However, starting at a probability level below 5%, the LI strategy had a higher NR at all probability levels.

With conventional market prices, the LI and OI 4-year strategies would be preferred to the 2-year HI strategy because they dominated the 2-year HI strategy by FSD and SSD, respectively. There was neither FSD nor SSD between the 4-year LI and 4-year OI strategies.

When the full organic premium for the OI strategy was added to the risk analysis under the 4-year sequence, the results of the OI strategy changed dramatically. The CDF of the OI strategy shifted notably to the right, and thus the OI strategy dominated the LI and HI strategies by FSD. The LI and HI strategies did not change since they were still receiving conventional prices, and thus there was still no FSD or SSD between the LI strategy and the HI strategy. Adding the full organic premium to the 4-year OI strategy

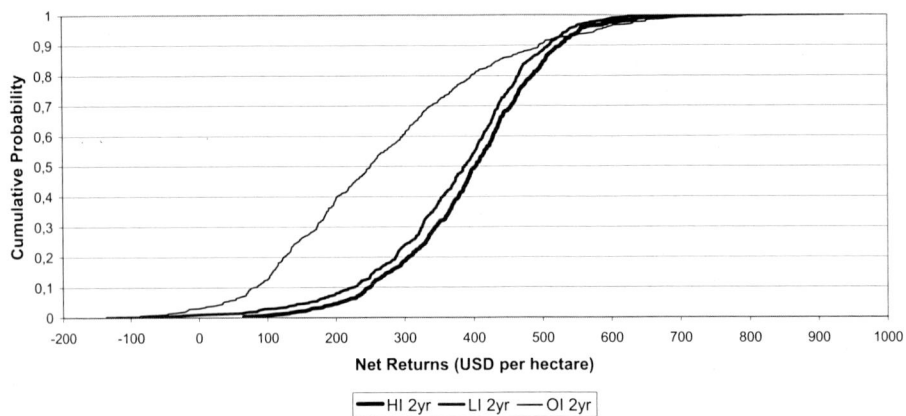

Figure 6. CDFs of Net Returns for LI, HI, and OI strategies, 2-year sequence with conventional market prices

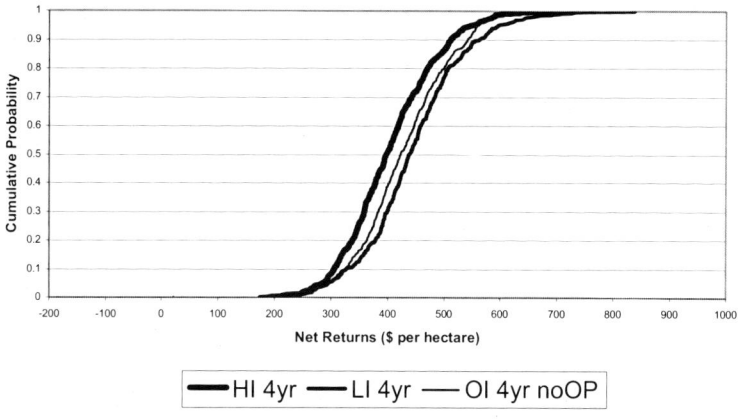

Figure 7. CDFs of Net Returns for LI, HI, and OI strategies, 4-year sequence with conventional market prices

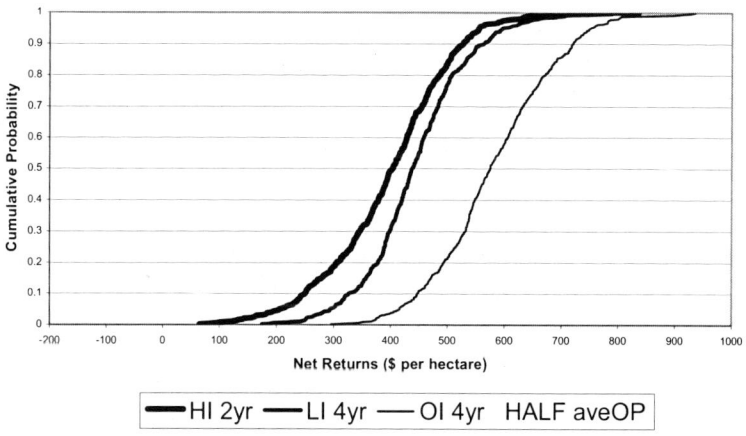

Figure 8. CDFs of Net Returns for HI-2yr and LI-4yr with conventional market prices and OI-4yr with half of the historic organic premiums

clearly made it the preferred strategy over all other input strategies and crop sequences.

Applying half of the historical average organic premiums to the OI crop in the 4-year sequence also resulted in the OI CDF moving notably to the right (Figure 8). The 4-year OI strategy was again clearly below and to the right, and thus preferred to both the LI 4-year and HI 2-year strategies when applying half the premium to the 4-year OI strategy. In other words, the 4-year OI strategy with half of the historical average organic premiums had a higher NR at all probability levels than the 2-year HI and 4-year LI strategies with conventional market prices.

6. CONCLUSIONS

This research has shown the long-term impact on net returns of organic cropping strategies compared to conventional strategies. Even though crop yields were lower under the 4-year OI strategy, so too were its production costs. As a result, the 4-year OI strategy was able to produce NRs per acre statistically equal to the NRs under LI and HI strategies without any organic premiums and significantly higher NRs when it received either full or half of the historical organic premiums. When the variability of NRs was analyzed using stochastic dominance, the 4-year OI strategy (with either full or half of the historical organic premiums) dominated all other strategies and sequences. When no organic premiums are applied, the 4-year LI strategy had higher NRs except at low probability levels (i.e., below 5%) although the CDFs are close visually. Based on these per-acre results, the original hypothesis of this study that conventional agriculture (as represented by the HI and LI strategies) was more profitable and/or involved less risk than a 4-year OI strategy must be rejected for this part of Minnesota.

This result that conventional agriculture is not obviously more profitable or less variable on a per acre basis supports the continuance of the current programs supporting organic farming such as production research (including crop insurance coverage), market information and development, and policies at the federal and state levels. Policies that include subsidizing farmers for the environmental benefits of organic production methods warrant further development and refinement. Companies in the food supply chain can continue to make investments to increase their capacity to handle organic products. Farmers and their advisors can be more confident in the potential benefits of investing the time and costs to learn the skills needed to grow and market certified organic products and to control potential problems.

Therefore, using the data from this study and the resulting profitability and risk analysis, the perception that conventional agriculture is more profitable and/or involves less risk than a 4-year, organic strategy, is not true for this part of southern Minnesota.

Further research, however, still is needed in several areas. This current study should be extended to include the VICMS trials that started on land with low fertility and high weed seed counts as well as the data from recent and future years. The NRs at the whole-farm level (versus the per-hectare calculations in the current study) should also be evaluated to estimate the impact of higher labor needs with organic methods. Further work is needed on the impact on crop prices and organic premiums of a substantial shift from the dominant corn-soybean sequence to longer cropping sequences. While considerable work has been done on the consumer demand for organic products, more work is needed on the entire supply chain for organic products. Research also is needed on the feasibility of smaller livestock production units on organic farms as well as the availability and associated costs of obtaining organic manure from larger livestock units. Additional research is needed on organic production methods including the feasibility of including green manures in the cropping system, organic weed control, and cropping sequences other than those included in VICMS.

7. REFERENCES

Dobbs, T.L., and Pourier, J.L., 1999, *Organic price premiums for northern Great Plains and Midwest crops: 1995 to 1998*. Economics Pamphlet 99-1, South Dakota State University, Brookings, SD.

Duram, L.A., 1998, Organic agriculture in the United States: Current status and future regulation, *Choices*, **13** (Second Quarter): 34–38.

Durgan, B.R., Gunsolus, J.L., Becker, R.L., and Dexter, A.G., 1992, *Cultural and chemical weed control in field crops-1992*, AG-BU-3157-S, Minnesota Extension Service, University of Minnesota, St. Paul, MN.

Fuller, E., Lazarus, B., Carrigan, L., and Green, G., 1992, *Minnesota farm machinery economic costs estimates for 1992*. AG-OF-2308-C, Minnesota Extension Service, University of Minnesota, St. Paul, MN.

Hardaker, J.B., Huirne, R.B.M., and Anderson, J.R., 1997, *Coping with Risk in Agriculture*. CAB International, New York, NY.

Krissoff, B., 1998, Emergence of U.S. organic agriculture—Can we compete?: Discussion. *American Journal of Agricultural Economics*, **80**(5): 1130–1133.

Minnesota Agricultural Statistics Service, 2001, *Minnesota Agricultural Statistics 2001*, USDA, National Agricultural Statistics Service, St. Paul, MN.

Olson, K.D., Weness, E.J., Talley, D.E., and Fales, P.A., 1992, *1991 Annual Report of the Southwestern Minnesota Farm Business Management Association*, Economic Report ER92-3, Department of Agricultural and Applied Economics, University of Minnesota, St. Paul, MN.

Porter, P.M., Huggins, D.R., Perillo, C.A., Quiring, S.R., and Crookston, R.K., 2003, Organic and other management strategies with two- and four-year crop rotations in Minnesota, *Agronomy Journal*, **95**: 233–244.

Roberts, W.S., and Swinton, S.M., 1996, Economic methods for comparing alternative crop production methods: A review of literature, *American Journal of Alternative Agriculture*, **11**(1):10–16.

United States Department of Agriculture, 2001, *Organic production and handling standards fact sheet*, Agricultural Marketing Service; www.ams.usda.gov/nop/facts/standards.htm (accessed Sept. 6, 2001).

Welsh, R., 1999, *The economics of organic grain and soybean production in the Midwestern United States*, Policy Studies Report No. 13, Henry A. Wallace Institute for Alternative Agriculture, Greenbelt, MD.

COMPARING THE PROFITABILITY OF ORGANIC AND INTEGRATED CROP MANAGEMENT
An analysis of apple and peach growing in Italy

Carlo Pirazzoli, Nicola Stanzani, Alessandro Palmieri, Roberta Centonze, and Maurizio Canavari[*]

SUMMARY

The present study, which is part of a broader research activity still under way, provides cost and profitability analysis with respect to organic and integrated fruit production in Italy. A comparison between the two production techniques within a static farm management framework is discussed. Production costs and profits are estimated for two case studies, the first one related to apple production and the second one related to peach and nectarine production. Both the cases refer to the area of Northeast Italy, which is the most important area for these types of produce.

1. INTRODUCTION

This study is part of a broader research activity of the Department of Agricultural Economics and Engineering. It provides cost and profitability analysis and consequently points out investment opportunities to farmers and other agricultural operators, with respect to the organic fruit production. This contribution draws heavily on a recent article published in Italian (Canavari *et al.*, 2004).

The research, comparing organic and integrated apple and peach production in Northeast Italy, was carried out considering the years 2002–2003. The aim of the study was to assess the investment gain in organic production without considering EU subsidies, which are not admitted for fruit growing in Italy.

The main steps of the analysis can be summarised as follows:

---

[*] Dipartimento di Economia e Ingegneria agrarie, Alma Mater Studiorum-Università di Bologna. Carlo Pirazzoli: sections 1 and 2. Nicola Stanzani, Alessandro Palmieri, and Roberta Centonze: section 3. Maurizio Canavari: section 4.

1. to estimate the production costs for integrated apple production, considering the existing agricultural system, production techniques, the main cultivars utilized in the area, and the average size of farms;
2. to estimate the production costs for organic apple production, considering the existing agricultural system, production techniques, the main cultivars utilized in the area, and the average size of farms;
3. to compare the two techniques with respect to production costs as well as to profitability.
4. to apply the same three-step process to peach production systems.

This paper focuses on the economic assessment, following the methodological path chosen. Experts from the realities investigated were interviewed for either technical or economic data collection.

2. MATERIALS AND METHODS

The research has been conducted using the case study method (Yin, 2002). Pilot-farms have been chosen for the analysis of the most common fruit growing and management techniques, in the area considered. However, the farms selected are not a statistically representative sample. The data have been collected directly on farms. Fruit growers and technical assistants have been interviewed; the production technique has been described; individual balance sheet items have been valuated; and the employment of resources for the scheduled activities have been identified.

The costs of production have been calculated for the normal crops of the area. According to a standard costing methodology, also adopted by Pirazzoli *et al.* (1999), three different economic aggregates have been identified: the *direct cost*, the *full cost* and the *total production cost*.

In the analysis of *direct cost*, the actual disbursement needed for obtaining the production has been considered. All the "out-of-pocket" expenses, related to the production process, have been calculated, while the structural expenses have not been considered. This estimate of direct costs provides useful insights through the comparison of different processes which apply the same production arrangement (cultivar or fruit species). The first step in calculating this estimate requires that the amount of raw materials used, the labor and the machine working hours, as well as the external services hired, be determined. The cost of each factor is calculated by multiplying the unit cost of the amount used. In particular, with respect to the labor, only the hired manpower is calculated. With respect to the machines, fuel and lubricant expenses as well as the maintenance charges specifically assigned to the processes examined are counted. The *direct cost* also includes the insurance premiums (anti-hail) and the plant depreciation, obtained by dividing the plant establishment expenses by the plant duration.

Full cost is obtained starting from the *direct cost*, to which the share of the common costs that can be attributed to the productive process considered (land maintenance and insurance charges, general and administration costs, welfare contributions, taxes, financial charges and rents) is added. In this specific case, it represents the overall amount of the costs attributed to the farm managed by an independent farmer. This is the most widespread form of estimation in the area considered, but it is not comparable with figures obtained for a different type of farmer.

Total production cost is the third and comprehensive measure. In order to compare farms that differ from each other in terms of relations between property, management and farming techniques, a homogeneous economic indicator is needed. Such an indicator considers all the costs related to a given productive process, including the opportunity costs for the farm's assets and labor.

Besides the items already listed, the full production cost also includes the unpaid labor (grower's and his family's manpower without social security expenses, which were already calculated), the land rent and the non-cash interest charged on the advanced capital (in this case an interest rate of 3% was adopted), on the plant expenses (obtained charging an interest rate of 2% over half of the plant and training expenses, *cf* Pirazzoli, 1993), and finally on the farm machinery.

3. RESULTS

3.1. Peach growing

To carry out the analysis of the peach production costs in the province of Forlì-Cesena the cultivar Redhaven, the most widespread type of peach in the area, has been used (Bertazzoli *et al.*, 1996). The average farm size is about 3 hectares (ha), managed by independent farmers and specializes in fruit growing. The planting arrangement is the *palmette*.

With respect to the integrated technique, an average production of 23 metric tons per hectare has been considered and a total production cost of about 11,500 EUR/ha was calculated, corresponding to 0.50 EUR/Kg (Table 1).

If the opportunity costs (unpaid labor, non-cash interest, etc.) are subtracted from the *total production cost* the *full cost* obtained is about 5,700 EUR/ha (0.25 EUR/Kg).

Such a value is constituted for about two thirds (almost 4,000 EUR/ha, 0.17 EUR/Kg) of *direct costs*: raw materials (8.3%), labor (11.6%), insurance premium (5.9%), and capital consumption costs (9.0%). The remaining 1,734 EUR/ha are *indirect costs* (land maintenance and insurance charges, general and administration costs, welfare contributions, taxes, financial charges and rents, etc.).

With respect to the organic production technique an average yield of about 19 t/ha and a total production cost slightly higher than 11,600 EUR/ha (0.62 EUR/Kg) have been estimated (Table 2).

Thus total and unit costs for organic production have been found higher than the costs for integrated production. The opportunity costs count for the 48.6% and the *full cost* is about 6,000 EUR/ha (0.32 EUR/Kg), of which less than 14% (that is 1,600 EUR/ha) are indirect costs.

Table 1. Production cost of integrated peach in the province of Forlì-Cesena (2002)[a]

ITEMS		EUR/ha	EUR/Kg	%
1.	Materials	948.78	0.04	8.3
2.	Manpower	1,331.83	0.06	11.6
3.	Hail insurance	676.20	0.03	5.9
4.	Plant depreciation	1,032.91	0.04	9.0
A.	DIRECT COST	3,989.72	0.17	34.7
5.	Common costs	1,734.00	0.08	15.1
B.	FULL COST	5,723.72	0.25	49.8
6.	Opportunity costs	5,770.03	0.25	50.2
C.	TOTAL PRODUCTION COST	11,493.76	0.50	100.0

a. Cultivar: Redhaven. Training system: palmette. Full production years: 10. Average production: 23 metri tons. Picking efficiency: 110 Kg/h.
Source: survey data.

Table 2. Production cost of organic peach in the province of Forlì-Cesena (2002)[a]

ITEMS		EUR/ha	EUR/Kg	%
1.	Materials	1,519.63	0.08	13.1
2.	Manpower	1,260.52	0.07	10.8
3.	Hail insurance	552.72	0.03	4.8
4.	Plant depreciation	1,032.91	0.05	8.9
A.	DIRECT COST	4,365.78	0.23	37.6
5.	Common costs	1,612.68	0.09	13.9
B.	FULL COST	5,978.46	0.32	51.4
6.	Opportunity costs	5,643.19	0.30	48.6
C.	TOTAL PRODUCTION COST	11,621.65	0.62	100.0

a. Cultivar: Redhaven. Training system: palmette. Full production years. 10. Average production: 18.8 metric tons. Picking efficiency: 95 Kg/h.
Source: survey data.

Among the *direct costs* (4,366 EUR/ha, that is 0.23 EUR/Kg), the raw material supplying costs share is 13% of the total costs (1,520 EUR/ha). The considerable difference between organic and integrated production technique (for which the raw materials count for just 8% of the total cost) is to be attributed to the higher expenses required for an effective plant protection in the organic production technique.

The items included in the *direct costs* are hired labor, which counts for 1,261 EUR/ha (10.8%), product insurance (about 553 EUR/ha, that is 4.8% of the total) and plantation consumption (1,033 EUR/ha that is 8.9%).

In computing the costs of organic cultivation, many items have a high level of uncertainty since the process is more dependent on the climatic conditions than integrated production. Similarly, the organic yield is subjected to seasonal trends.

Analyzing the variation of the production costs for cultivars other than Redhaven, the same economic items have been considered. In particular, *Springcrest* (early common peach), *Fayette* (late common peach), Rita Star (nectarine) and *Stark Red Gold* (nectarine).

With respect to total costs, it increases going from early to late varieties both for peaches and nectarines. This is due to higher harvesting as well as higher phytosanitary

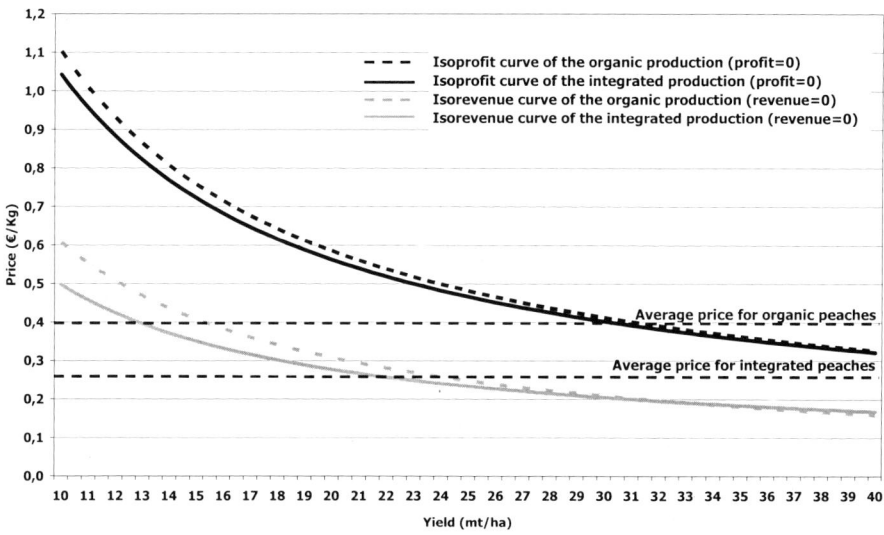

Figure 1. Isoprofit and isorevenue curves for integrated and organic peach growing

treatment expenses (necessary for a longer period of time for late cultivars) and cash outlays for the production insurance. The incidence of the pruning costs differ, it is lower for late cultivars than for early ones (Canavari et al., 2004).

Thus, the costs recorded for organic production are higher than the costs of integrated fruit production with respect to early cultivars and lower with respect to late ones. Indeed, adapting the integrated production technique, late cultivars require higher investment in fruit picking.

In general, a lower incidence of opportunity costs is found for late varieties.

If the costs per unit of production are considered, these show a trend opposite to the costs per hectare, decreasing from early to late varieties. In this case, the cash outlays for organic production are always higher than those for integrated production (with differentials of at least 20% for early cultivars).

In order to appraise the profitability of peach production, the costs of production have been compared to the market prices using the isoprofit and the isorevenue curves (Figure 1). These lines represent a set of points in output space that all yield the same profit or revenue (Bowles, 2004). In this specific case, the isoprofit curve shows all the combinations of yields and market prices that nullify the farmer's profit, considering the total production cost already estimated. The isorevenue curve shows all the combinations of yields and market prices that nullify the farmer's revenue, considering the calculated full cost (Bertazzoli et al., 1994).

It is noted that for integrated production, considering a yield of 23 t/ha, the minimum price that allow the farmer to get a profit is 0.25 EUR/Kg. This is in line with the prevalent price of the most important farmers' cooperative in the area (Apofruit) for the period 1999–2002 (Stanzani et al., 2002a and 2002b).

For organic farming, given a production of about 19 t/ha, a price of at least 0.32 EUR/Kg is required to get a positive revenue. This price is considerably lower than the prevalent one recorded in the Apofruit price lists of the latest years (0.40 EUR/Kg).

3.2. Apple growing

To analyze the apple production cost in the suitable agricultural areas of the Trentino-Alto Adige Region, the cultivar Golden Delicious was used for both the farms adopting integrated crop management (in the Trentino sub-area) or an organic crop management (in the Alto Adige sub-area). As with peach growing in Forlì-Cesena province, the apple farms are mostly operated by independent farmers and have an average size of 3–5 hectares (Bacarella and Schifani, 1996). The training system is the Spindlebush.

In the plantations grown following integrated production techniques, the average yield is about 50 t/ha and the total production cost is estimated at 16,000 EUR/ha, that is 0.32 EUR/Kg (Table 3).

The opportunity costs are 40% of the *total production cost*. Thus, the *full cost* is about 9,707 EUR/ha (0.19 EUR/Kg). *Direct costs* account for about 50% (4,995 EUR/ha that is 0.10 EUR/Kg) of the total production cost and includes raw material costs (10.6% of the total cost), labor (5.4%), hired services (5%) and plantation depreciation (10.1%). The remaining 4,711 EUR/ha are *common costs*. Looking at the organic production, the yield was estimated at 35 t/ha and the total cost of production at about 17,000 EUR/ha (0.49 EUR/Kg) (Table 4).

Thus, the unit cost for organic production is estimated to be more than 50% higher than that for integrated production. In this case, the opportunity costs account for slightly more than 30% and the full cost is 11,755 EUR/ha (0.34 EUR/Kg), of which less than 30% (4,932 EUR/ha) represents common costs.

The lower incidence of opportunity costs for integrated production (Trento province) has to be attributed to the high quality of the family labor, which allows a high output at harvesting (Canavari et al., 2004).

For direct costs (6,822.93 EUR/ha, that is 0.19 EUR/Kg), raw materials account for about 10% of the total cost (1,707.56 EUR/ha), while the hired labor represents 502.11 EUR/ha (3%). Quite significant is the product insurance premium (2,625 EUR/Kg, that constitutes 15.4% of the total) as well as the plantation depreciation (1,988.25 EUR/Kg that is 11.7%).

Besides the uncertainties related to climatic trends that are recorded for organic production, the costs for integrated and organic production techniques are difficult to compare due to differences between management strategies in the farms considered. In particular, no external services are hired by the farmer producing organically. On the other hand, no product insurance is paid by the farmers that grow apples following the integrated technique.

The apple production costs have been compared to the market prices using the iso-profit and the isorevenue curves, just like the technique used for peaches. In the case of integrated production, given a yield of 50 t/ha, a profit can be gained by the farmer when considering a selling price of about 0.35 EUR/Kg (Figure 2). On the other hand, given an average organic production of 35 t/ha, a profit can be gained starting from a market price

Table 3. Production cost of integrated apple in the province of Trento (2002)[a]

ITEMS	EUR/ha	EUR/Kg	%
1. Materials	1,709.02	0.03	10.6
2. Manpower	860.76	0.02	5.4
3. Hail insurance	797.96	0.02	5.0
4. Plant depreciation	1,627.58	0.03	10.1
A. DIRECT COST	4,995.32	0.10	31.1
5. Common costs	4,711.86	0.09	29.3
B. FULL COST	9,707.18	0.19	60.4
6. Opportunity costs	6,363.51	0.13	39.6
C. TOTAL PRODUCTION COST	16,070.69	0.32	100.0

a. Cultivar: Golden Delicious. Training system: Spindlebush. Full production years: 17. Average production: 50 metric tons. Picking efficiency: 150 Kg/h.
Source: survey data.

Table 4. Production cost of organic apple in the province of Bolzano (2002)[a]

ITEMS	EUR/ha	EUR/Kg	%
1. Materials	1,707.56	0.05	10.0
2. Manpower	502.11	0.01	3.0
3. Hail insurance	2,625.00	0.08	15.4
4. Plant depreciation	1,988.25	0.06	11.7
A. DIRECT COST	6,822.93	0.19	40.1
5. Common costs	4,932.42	0.14	29.0
B. FULL COST	11,755.35	0.34	69.1
6. Opportunity costs	5,245.36	0.15	30.9
C. TOTAL PRODUCTION COST	17,000.71	0.49	100.0

a. Cultivar: Golden Delicious. Training system: Spindlebush. Full production years: 25. Average production: 35 metric tons. Picking efficiency: 225 Kg/h.
Source: survey data.

of 0.50 EUR/Kg. Such prices are consistent with or even lower than the prices recorded in the latest years, for both integrated and organic production.

4. FINAL REMARKS

Comparing organic and integrated peach growing in the selected area (Forlì-Cesena province), the present analysis pointed out a considerable difference in the economic performance. The appreciation of organic products on the market is from 30% to 90% higher than that for integrated products. This difference largely compensates either the small cost increase or production decrease recorded for organic productions. The selling prices highly influence the total economic results. Thus, the variety selection should be oriented towards the more productive late fruit *cultivars* that allow a lower production cost per unit.

Apple production in the Trentino-Alto Adige Region (which entails the provinces of Trento and Bolzano), for either integrated or organic, seems to ensure good revenues.

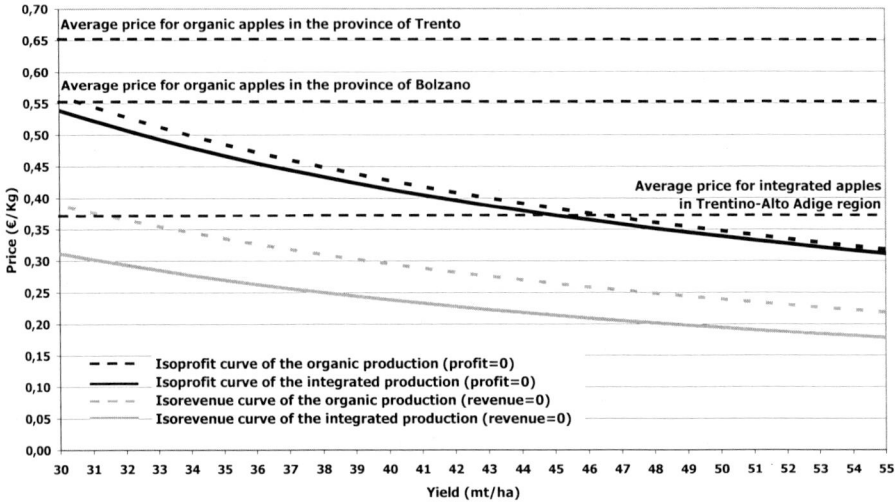

Figure 2. Isoprofit and isorevenue curves for integrated and organic apple growing

Thanks to recent strong incentives to the adoption of organic agriculture promoted by the government of the two authonomous provinces, the organic apple plantations are widespread in the Region but they contribute only for 4% of the produce marketed by farmers' co-operatives. Indeed, the co-ops are well oriented to the promotion of integrated grown products in both the Italian or foreign markets. Moreover, the many technical difficulties for organic production in the Trento province could justify such a limited diffusion in the area. In the province of Bolzano, larger quantities of products and quotation prices lower than those in the Trento province are recorded. In particular, the price difference between organic and integrated products from Alto Adige is about 10%.

In conclusion, the present economic and technical conditions (in terms of yields and prices) make the conversion from conventional to organic peach production convenient in the Forlì-Cesena province. However, integrated and organic apple production in Trentino-Alto Adige does not show significant differences in profitability. These conclusions come with the warning that it was not possible to compare study cases that were homogeneous at a province scale.

5. REFERENCES

Bacarella, A., and Schifani, C., 1996, La frutticoltura in Italia, in: *La competitività della frutticoltura italiana*, F. Alvisi, A. Bacarella, P. De Castro, eds., Nomisma, Bologna, pp. 1–50.

Bertazzoli, A., Pirazzoli, C., Malagoli, C., and Regazzi, D., 1994, Gli aspetti economici, in: *La filiera ortofrutticola in Emilia-Romagna*, F. Alvisi, D. Regazzi, eds., GeSTA-CNR, Bologna, pp. 125–202.

Bertazzoli, A., Pirazzoli, C., Malagoli, C., and Regazzi, D., ,1996, Gli aspetti economici della fase agricola, in: *La competitività della frutticoltura italiana*, F. Alvisi, A. Bacarella, P. De Castro, eds., Nomisma, Bologna, pp. 51–108.

Bowles, S., 2004, *Microeconomics: behavior, institutions and evolution*, Princeton University Press, Princeton.

Canavari, M., Pirazzoli, C., and Stanzani, N., 2004, Analisi di costi e redditività in aziende frutticole biologiche, *Rivista di Frutticoltura*, **2**: 35–39.

Pirazzoli, C., 1993, Metodologia di calcolo dei costi di produzione, in: Proceedings of the Conference *Riduzione dei costi di produzione in orto-floro-frutticoltura*, Faenza, October 8, 1993, SOI-Società Orticola Italiana, Firenze.

Pirazzoli, C., 1999, Sistemi produttivi a confronto: aspetti metodologici, in: *La peschicoltura nell'Unione Europea: comparazione economica tra i principali sistemi produttivi*, CSO-Centro Servizi Ortofrutticoli, Ferrara, pp. 10–17.

Stanzani, N., Aldini, A., and Valdinoci, G., 2002a, Integrata, profitti negativi. I prezzi premiano il biologico, *Terra e Vita* (suppl.), **35**: 24–30.

Stanzani, N., Aldini, A., and Valdinoci, G., 2002b, Integrato e biologico: i costi di produzione, *Agricoltura*, **6**: 24–27.

Yin, R.K., 2002, *Case study research. Design and methods*, Sage publications, Thousand Oaks.

MARKETING STRATEGIES FOR ORGANIC WINE GROWERS IN THE VENETO REGION

Luca Rossetto[*]

SUMMARY

Recently, the Italian organic wine sector has dramatically increased. In the last two years, the organic vineyard area has doubled, reaching more than 40,000 hectares, while organic wineries account more than 9,000 farms. In particular, the Veneto Region accounts for 4% of total area and for 15% of organic wine makers.

This study analyzes organic wine sector in the Veneto Region, mainly focusing on marketing issues, through a questionnaire survey.

Two main organic wine enterprises have been recognized: small wine growers, specialized in producing organic grapes and large-size wineries, highly specialized in both cultivating grapes and processing wine. Almost all the production is certified and highly differentiated; most of the product is sold on foreign markets, while domestic consumption is limited.

Two main marketing strategies have been detected among small vine growers and wineries. While big size wineries are focused on price and product variety, small vine growers, selling wine to traditional retailing, follow wine quality strategy. Indeed, big producers selling to foreign or domestic supermarket chains or to final consumers have high market power inside the wine food chain.

Actually, Italy is still one of the leading organic producers in the world, also the quality of its product is perceived as having high standard. However, lack of well designed marketing plan could be a weakness for Italian exporters, as increasing competition from laggard countries, such as US, Canada, Australia, whose companies draw a well planned marketing strategy, may affect the Italian overseas market.

---

[*] Associate Professor of Agricultural Economics at Dept. TeSAF, University of Padova, Agripolis Via Romea, 35020 LEGNARO (PD) Italy; e-mail: luca.rossetto@unipd.it.

1. WINE FROM ORGANIC AGRICULTURE

So far, organic wine is defined world-wide as "wine made from organically grown grapes". The fundamental idea behind organic wine is that making wine from grapes grown without chemical fertilizers, weed-killers, insecticides and other synthetic chemicals is better both for ecosystem and wine consumers.

Actually, the US definition for "organic wine" is still pending USDA approval, while within EU the EC Regulation 2092/91 recognizes raw agricultural products as organic (grapes) excluding processed ones (wine). Meanwhile, self-standards for "organic wine" have been voluntarily adopted by certifiers and wine-makers. These standards often limit the use of sulfur dioxide, sugar adjustments, and the use of yeasts. In some instances, they provide regulation for acid correction, filtration/fining methods or storage in wood or metal barrels.

The use of sulfur dioxide has so far been debatable. The practice of adding sulfites to wine for protection against oxidation and bacterial spoilage use to be considered harmful for consumers' health because it might cause headaches and allergic reactions. Actually, these results happen when the use of sulfites is excessive[1].

Recently, the US National Organic Standards Board (NOSB) has overturned their previous judgment about the use of sulfites. The new National Organic Program (NOP) recommends sulfur dioxide in wine labeled "made with organic grapes" because prohibiting its use would have a negative impact on grape production and wineries, but the sulfite content must not exceed 100 parts per million. The use of sulfur dioxide in organic wine is also admitted by IFOAM[2] (IFOAM, 2000) and Codex Alimentarius Commission[3] (Fao/WHO, 2001).

Notwithstanding, many producers are rallying for regulations of the wine-making process, not only in Europe or in the US but all over the world. Unlike this harmonization, the German, French, American, Japanese or Italian consumer may drink wine made from organic grapes but the product can be very different in such things as sulfite content, storage in wood or steel or plastic barrels, etc..

Actually, three different kind of wines can be found on the Italian organic market: a) "wines made from organically grown grapes" without any sulfite or additives; b) "wines made from organically grown grapes" processed as conventional wines; c)"wines made from organically grown grapes" where processing as well as sulfite and additives are defined and controlled by certification agencies. The three categories cannot be easily recognized by consumer since ingredients or processing cannot be reported on the label.

The lack in regulating organic wine-making process has encouraged producers to inform their consumers directly on product features according to their own disciplinary procedures (Pinton, 2001). In other cases, wine-makers verify their own suppliers by checking both the production and the wine-making process.

Though, the harmonization of different national regulations seems far away. In the EU, where the organic wine market "is going fast out of the niche size" (Compagnoni,

---

[1] The excess in sulfites use is often found in white wines or wine made from spoiled grapes (ex. grapevine spoiled by fungi diseases).
[2] The approved ingredients of non agricultural origin and processing aids used in food processing are listed in Appendix 4 of Basic Standards for Organic production.
[3] In spirituous beverages with less than 15% alcohol, the max level of sulfites (as residual) is 150 mg/kg.

2001), EC Reg. 2092/91 lacks content about the wine-making process.[4] This put together with labeling issues creates difficulties for market development and promotion of organic wines. Some points may explain this issue:

- the use of term "organic wine" is forbidden[5];
- the term "wine made from organically grown grapes" is authorized on quality wines produced in specific regions (VQPRD) and wines with geographic specificity, but it is not allowed on table wines or products made form them (ex. vinegar);
- the use of the EU organic logo is not allowed;
- in Italy, information on specific production practices cannot be reported together with compulsory data[6] on degree proof, volume, parcel number, etc.

As long as organic wines are not well-defined it will be very difficult to organize the promotion to consumers.

Technical and agronomical aspects are also important issues. Organic farming can easily be implemented in vineyards. However, a main concern is the issue of soil cultivation, which may imply the use of more labor and skilled work. Still, there is a major problem concerning the impact of copper use on soil. In fact, the copper salts are efficient fungicides against downy mildew (*Plasmopara viticola*), grapevine powdery mildew (oidium, *Uncinula necator*) or Botrytis (*Botrytis cinerea*) and it is allowed in organic farming. Since copper salts slowly accumulate in the soil, wine-growers try to reduce, as much as possible, their use but "ecological" or "natural" effective alternatives do not exist, while the EU regulation fixed in 2002 the deadline for copper use. Recently, the restricting EU regulation on copper use has been updated by EC reg. 473/2002 which fixes the amount of copper per hectare to 8 kg through the year 2005 and 6 kg there after. This regulation appears to be less restrictive than the previous one, which makes wine-growers less worried about the future.

2. OVERVIEW OF THE ORGANIC WINE MARKET

Statistics on world organic vineyards are not uniform for both years and sources (Yussefi and Willer, 2004).

In North America organic vineyards are recognized in California; in Latin America there are organic areas in Argentina and Chile; organic wine productions are also reported in Australia (FAS, 2004; Senasa, 2004). Organic grape production in California occurs in nearly all regions, and the total area managed as organic is approximately 4,870 hectares (ha) (Greene, 2001). Most of Californian vine-growers have adopted soft chemistry diseases depending on climatic conditions, insect and disease pressure. The growth of the organic vineyard is mainly observed in all Mediterranean countries but the production reaches a critical mass in Italy, France and Spain (Table 1).

---

[4] In the EC reg. 2092/91, Annex VI, many usual additives such as bentonite, casein are reported while sulfur dioxine additive not mentioned.
[5] Even if "wine made from organic grapes" is allowed by EU regulation only, "organic wine" is found on some UK bottle label as well "Ökologisch wein" on German bottle.
[6] Art. 17, EC reg. 3201/90.

Table 1. Organic vineyards in wine producer countries in 2000

	Organic vineyards Total area (ha)	Share % on total vineyard area	Share % on total organic area
Austria[a]	680	1.3	0.2
Germany[a]	2,260	2.3	0.5
Spain[a]	15,470	1.5	2.2
Greece[a]	450	0.5	6.6
France[a]	13,250	1.5	0.4
Italy (2001)[b]	44,175	0.8	3.6
Portugal[a]	2,780	1.3	2.4
Argentina[c]	6,500	0.4	16.6
USA[d]	5,100	0.5	0.6

Source: [a] Häring et al., 2004; [b] Ismea, 2004; [c] OIA (Organizacion Internacional Agropecuaria) 2003; [d] Economic Research Service, 2004.

After these countries, Portugal has a significant organic vineyard area, while Austria[7] and Greece reach 700 and 550 hectares, respectively.

A fast growing production of organic grapes is observed in France, where the domestic market still suffers an ambiguous regulation on the wine-making process and consumer's concerns about the quality of organic wines. Consequently, this causes the market growth forecasts to be lower than other organic produce. French supermarket consumers believe that the organic wine price cannot be higher than the conventional one: therefore, French organic wine-makers want to produce high quality wine to get prices similar to this market segment (Gardner, 2000).

Germany represents the main organic market, even for organic wine (Haccius and Lünzer, 2001). This country mostly produces high quality wines despite the fact that their share on total wine consumption is not greater than 1%. Yet, the production cost of organic wine is 30% greater than conventional wine because of lower yields and higher input expenses. The supply shortage requires imports from others countries, especially Italy and France, which amounts to about 70% of total supply, i.e. 55% of conventional wine market. Main marketing channels are from winemaker to final consumer (30% of total sales), restaurants (15%), traditional wholesalers (15%) and specialized bio-shops (15%). A low share of organic wine production is also sold in the UK and internet sales are increasing. The German organic wine is expanding from its niche position, which is promoted in part to a marketing strategy that emphasizes the quality of organic wine (Kopfer and Willer, 2001).

3. THE ORGANIC WINE MARKET IN ITALY

During the 1990s many enterprises entered the segment of "wine made from organically grown grapes" at all supply-chain steps: production (farms), processing (cooperatives) or industry (Castaldi, 2000).

---

[7] Austrian farms cultivating organic vineyards amounted to only 1% related to the total.

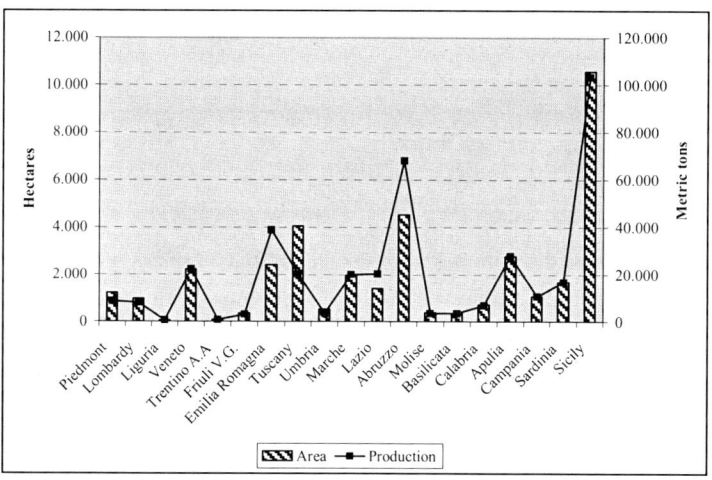

Figure 1. Vineyard area and grape production per Italian region (Ismea, 2004)

In the beginning of 2003, the area of organic-grape oriented farms was extended to 37,379 hectares, and was located mostly in three regions: Sicily (28%), Abruzzo (12%), and Tuscany (11%). A significant share of organic-grape area, between 5 and 7%, is located in Emilia-Romagna, Marche and Apulia (Figure 1) (Lunati, 2002; Lunati, 2004; Mingozzi and Bertino, 2005; Santucci and Pignataro., 2002). Also in the northern regions, the organic viticulture growth has been remarkable, (but lower than in Southern Italy), especially in Veneto Region. In 2002, the share of organic vineyards in Veneto Region was 6% over the national organic total area. Organic grape production follows the same pattern area except in the Emilia Romagna, Abruzzo and Lazio regions where productivity is a little bit higher than the Italian average (10 metric tons per hectare).

Italian farms producing wine from organic agriculture vary from small to medium and big business size. There are farms cultivating less than one hectare and big scale farms managing more than 200 hectares. On the average, organic farms show an area greater than conventional ones, while farmers are definitively younger than the national average (Sgarbi, 2001).

Although available data are not updated so far, the positive trend in wine made from organic grape[8] is unambiguously positive and opposite to the general decline in wine per-capita consumption. This evolution shows that many organic wines have been graded high quality, i.e., these wines are mostly produced in specific regions where farmers, organized in Consortium and following measures, certify their products as DOC or DOCG[9].

---

[8] In 1998, the production of wine from organic agriculture was estimated in one million hectoliters (Sgarbi, 2001).

[9] The Italian terms D.O.C. (*Denominazione di Origine Controllata* or "Controlled Denomination of Origin") and D.O.C.G. (*Denominazione di Origine Controllata e Garantita* or "Controlled and Guaranteed Denomination of Origin") are indicative of two certification processes which guarantee that wines and grapes have been produced in specific areas following methods and practices set by the Consortium (e.g., yields are limited, production area is marked off, etc.).

Although the demand for wine made from organic grapes is still low related to other organic productions, the price is often not satisfactory. Organic wine consumption may grow at a lower rate than one observed in overall organic products, because the standard "organic" consumer seems less inclined to drink wine (Castaldi, 2000)[10].

However, organic wine exports show strong growth, enjoying the positive trend in consumption of high quality wines in many foreign countries (ISMEA, 2002). This flow of trade has enhanced the value of organic wine making many organic pioneer farmers successfully. In fact, many organic farmers export a large share of wine (55% of the total) in comparison to non-organic farmers (Sgarbi, 2001). Countries where the Italian organic wine is well-appreciated are Germany, Switzerland, UK and the US.

In the domestic market, organic shops are less interested in wine than the large-scale retail, because the consumer accustomed to organic shops, generally shows a low propensity in drinking alcoholic beverages.

Currently, the sector of wine made from organic grapes seems to be at a standstill, waiting for a EU regulation of wine-making process to give a clear definition of organic wine since a reasonable share of wine is currently sold in conventional markets.

Actually, experiences carried out in both cultivation and marketing show that the organic wine success is strictly linked to the production of high quality wines (Didero, 2001)[11].

4. A SURVEY ON ORGANIC WINE MARKET IN THE VENETO REGION

In this section, results of a sample survey of wine-growers in Veneto Region are presented. In particular, a sample of wine-makers, located in Veneto Region, has been analyzed by investigating farm features, management, supply, technology employed and the market. The study on the market has been further explored focusing attention on marketing strategies.

The Veneto Region is the most important for wine production in Italy. In 2001, the total vineyard area in Veneto Region amounted to 73,780 ha (ISTAT, Census 2001); farms cultivating grapes were 77,191 ha while the wine production reached about 8.8 million hectoliters and the wine market accounted to about 11 million EUR (11% of total agricultural sales).

According to 2001 data on the agro-environmental program (EC Reg. 1257/1999), the Veneto Region had 242 organic wine-growers with 2,047 ha of total agricultural land and 1,277 ha of vineyard area[12]. The share of organic farms with vineyards was around 9.7% and 20%, respectively, of total organic area and farms[13]. Due to its small size, the sample has been extracted from the universe of organic farms cultivating grapes on the

---

[10] This is in contrast with recent researches showing, especially for red wines, a link between some features and positive effects on health.
[11] Recently, the international competition of organic wines "Wine Award" held at the Biofach in Norimberga (Germany) was won by Italy. In particular the first place was reached, at equal merit, by two red wines from Veneto Region.
[12] These data also include small organic farms where vineyard is less than 1 hectare. However, these data represent only a sample of the universe.
[13] In 2000, organic farms in Veneto Region where 1,270 with 13,092 hectares (Biobank, 2001).

basis of their territorial (and administrative) distribution and economic features (production, wine-making and marketing holdings) instead of stratification. Actually, a sample of 32 farms was selected after a screening survey, which showed farmers' willingness in participate in the survey.

These farms are located in all Veneto Provinces where vineyard area is spread and wine production is definitively the most important agricultural product: Verona (hilly area), Vicenza (Berici hills), Padua (Euganei hills), Treviso (Montello and Conegliano area) and Venice (northeast area) (Figure 2).

The survey was carried out through the process of farmers filling out a questionnaire during the summer of 2001. The questionnaire is divided into three sections in order to investigate the farm structure, the management and supply-chain relationships. The analysis shows aspects of innovations, supply and market issues, strengths and threats of organic wine-growers according to their production and marketing strategies.

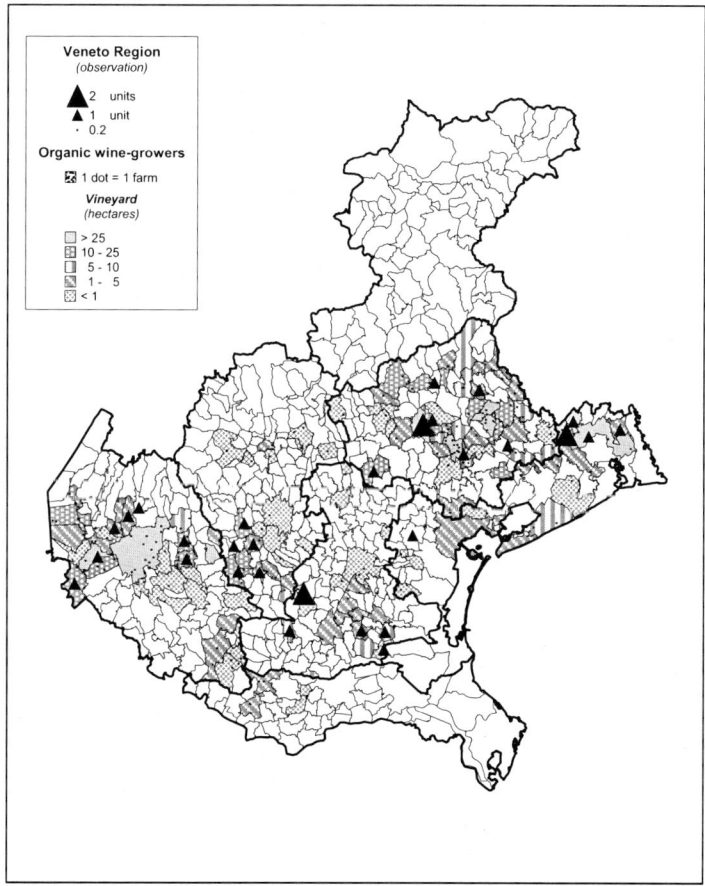

Figure 2. Geographic distribution of organic wine-growers in Veneto Region

4.1. The farm structure

Most of the farms in the survey produce grapes and make wine. They are also highly specialized in both cultivating the vineyard and processing wine and the organic agriculture practices are well-established. The size of the average sample farm is greater than the regional average: the agricultural farm area is around 13 hectares, of which nearly half is vineyard.

In Table 2 the farm distribution by type of faming—only grape, grape & wine, only wine, mixed farming[14]—is reported showing number, agricultural area and vineyard area. In mixed farming most of the land is cultivated to crops other than vineyard, while holdings of only wine are mainly cooperatives.

Almost two-thirds of total land is managed as organic agriculture (less than 2% of total land is in conversio[15]) while 94% of vineyard is cultivated with organic agricultural methods. The distribution of holdings per size class of vineyard area shows two types of organic farms:

1) small-size farm with an average vineyard of 4–7 ha. This one describes 60% of total farms with 22% of total organic vineyard;
2) large-size farms with more than 24 ha per unit, representing 40% of the sample and 78% of total vineyard[16].

4.2. The management

A critical aspect in converting a farm from conventional to organic farming is the management of human resource availability and skill. Manpower requirements are mainly satisfied by the farmer and his/her family as it is observed in overall Italian agriculture. Specifically, the family manpower satisfies 80% of total requirements, especially in cultivating vineyard and processing wine, but excluding harvesting and the sparkling process.

Most of the farms have converted a previous conventional activity and started this process a long time ago. Many farms decided to enter organic agriculture in order to exploit markets with greater potential profits thus increasing the value of professional experience (Table 3).

Managerial skills in organic farming, considered high especially in processing wine, are satisfied by the farmer and his family. Often pruning, pest management control, wine-making processes and marketing procedures are assisted by skilled operators from outside the farm.

Management training issues are also investigated focusing attention on technical, legal and marketing aspects.

Technical management training is often driven by self-experimentation (46.7% of holdings). The wine-making process information is acquired through professional advice, specialized press, or participating in seminars and exhibitions.

---

[14] The mixed farming includes vegetable crops and trees.
[15] According to the EC reg. 2092/91 the farm becomes "organic" after two years conversion period.
[16] This distinction does not include cooperatives which vineyard area referring to each coop-members was not declared.

Table 2. Number and area per farm by type of farming

	Type of farm				Total
	only wine	only grapes	wine/grape	mixed	
Number	3	8	16	5	32
Total area (ha)	23.3	7.2	11.6	22.0	13.4
Agricultural area (ha)	6.7	7.3	10.4	18.5	10.6
Vineyard area*	6.7	5.0	9.9	4.1	7.5

* Including both organic and conventional area.
Source: Defrancesco et al., 2002

Table 3. Information on organic farming activity (in %)

	Type of holding			
	only grapes	grapes/wine	only wine	marketing
General information:				
– already started activity	90.0	68.8	50.0	100.0
– new activity	10.0	31.3	50.0	
When did the activity start?	1975	1946	1991	1970
Why (reasons):				
– profitability	20.0	12.5		100.0
– professional skills	60.0	43.8		
– profitability and prof. skills	20.0	18.8	50.0	
– others		25.0	50.0	

Source: Defrancesco et al., 2002.

For cultivating organic grapes, information comes from producer unions or agricultural union party assistance. Certifying bodies play a crucial role in training management in all aspects of the organic wine supply chain, thus not only helping operators on legal aspects but also on enhancing professional skills. The information on legal issues is given mainly by certifying bodies and also by producer unions, especially in only-wine ones.

The critical point about information concerns marketing strategies. Only a limited share of producers feel investing in human capital is a priority. They try to meet this demand through direct customer relationships than making a marketing plan based on a specific market analysis.

Thus, organic wine-growers approach to the market has emerged as a weakness, especially in the upper stream supply-chain. Actually, organic wine-growers should change their purposes according to the market growth:

- in the infant and introduction growth stages of organic wine life cycle, the producer increases his professional skills in both technical and legal aspects of production regardless of market size;
- in the growth stage, the producer should make up adequate marketing and communication strategies because of increasing competitiveness among farms or processing companies, especially when market size reaches the maturity stage.

4.3. Economic results

Even if this survey was not calibrated on organic farm profitability, some economic aspects have been figured out. In particular, the growth rate of sales is high as in other advanced business. The annual sales has grown 46% in two years, reaching in the year 2000 a value of 24,000 EUR in small size farms (<5 ha), 60–70,000 EUR in medium size farms (10–15 ha) and 370,000 EUR in big size farms (>15 ha).

The direct cost of organic grape production is around 2,700 EUR per hectare. Costs are mostly made up of harvesting (28.6%), pruning (24.1%) and pest control (23.7%) and they can range according to mechanization and manpower costs. The organic production cost of vineyard is not far from the conventional one because the reduction in pest control costs (–45%) is slightly offset by an increase in other costs, especially harvesting costs (+51%).

The direct cost of organic wine processing is around 1.34 EUR per bottle for table wine, 2.12 EUR per bottle for sparkling wine and 3.87 EUR for wine matured in *barrique*. The organic wine cost varies according to the processing technology employed. The certifying cost must also be included, and it is on average equal to 97.6 EUR per hectare of vineyard.

4.4. The organic wine supply: vineyard and technology

The organic wine supply is influenced not only by agronomic and technical aspects but also by the market performance. The high yield variability faced by organic wine-growers is rarely counterbalanced by high market prices. Since a reduction in production may lead to low farm incomes, wine-growers adopt two main strategies: 1) small-size farms give up the wine-making process; 2) medium-big farms increase the production of high quality wines.

The survey then explored the vineyard structure focusing attention on yield instability and wine quality. There are five vineyard varieties of red grapes—Cabernet Franc, Cabernet Sauvignon, Marzemino, Merlot, Refosco—ten of white grapes—Chardonnay, Pinot grigio, Pinot bianco, Prosecco, Riesling, Sauvignon, Trebbiano, Garganega, Tocai, Verduzzo—and other local varieties (Incrocio Manzoni, Verduzzo friulano, etc). The average vineyard is about 3 hectares. All vineyards are quite old (18 years), because most were previously cultivated through conventional farming methods. The average yield is around 10 t/ha but this is fairly uncertain because it can lower to 3.8–4.0 t/ha due to unfavorable climatic conditions.

The grape price changes according to vineyard variety: from 0.3 EUR/kg in Tocai grapes to 0.5–0.6 EUR/kg in Caberbet ones or 0.7–0.8 EUR/kg in Pinot grigio. The grape price volatility is also high and strongly influenced by the market performance. Usually, the share of grapes directly sold to processors is positively linked to the grape price and negatively to farm-size (Figure 3).

The farm size strongly affects the technology, especially in the wine-making process. The survey has explored technology employed by organic wine-makers focusing not only on machinery but also its professional requirements. Processing wine technology employed by organic wine-makers does not significantly differ from technology employed by non-organic ones. Actually, larger organic farms have adopted a sophisticated technology in both cultivating vineyard and processing wine.

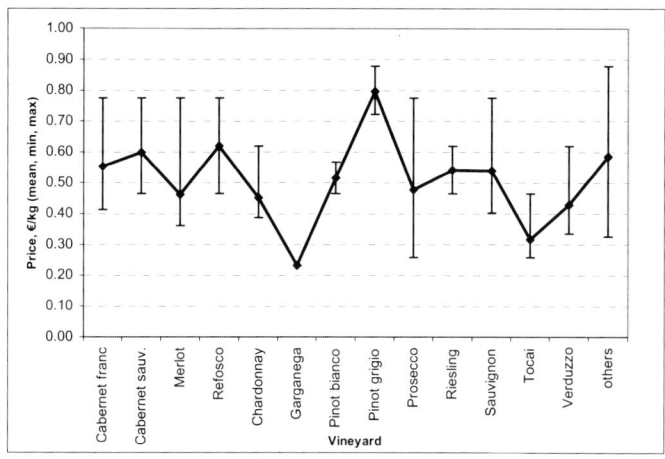

Figure 3. Price of the grape per vineyard, in EUR/Kg (Defrancesco *et al.*, 2002)

Finally, the evolution in production of high quality wines (DOC or DOCG) has been analyzed.

Wine-growers have gradually increased the production of organic DOC wines as long as it increases consumer perception about quality, and it reduces marketing risk, i.e., the organic wine can be allocated easily within the market.

For almost 40% of organic wine-makers the supply does not meet the market demand. Yet, perspectives for a further supply growth is around 2–3% per year. Many wine-makers in Veneto Region have also reached such a production size to meet supermarket demand.

4.5. The market

Organic wine-growers exhibit an explicit market attention by selling their production where they can get a greater value. Generally, organic wine is mostly sold to wholesalers and, more recently, to the large scale retail system (LSR). While the share of production sold directly to consumers is higher in sparkling or matured wines because of their greater value added and difficulties in breaking into other channel such as the LSR.

The sale distribution for geographic area is strongly affected by demand fragmentation. Actually, the market for organic wine is still in a starting growth stage: the market demand has a niche size while the consumption comes from few buyers scattered in many areas. In this scenario, extending the market may be the most convenient strategy in order to catch the opportunities of a fast growing market and, at the same time, to consolidate profitable market positions. Wine-growers seem to be able to catch this opportunity through sale diversification in the regional, national and foreign market, especially for regular wines, while low volumes of sparkling and aged wines limit their supply only to local markets.

As it always happens in the conventional wine market, the bottling process is increasing in the organic market because it improves the value-added while consolidating the image of the organic wine and wine-maker among consumers particularly aware of quality and naturalness.

The market analysis is focused on the relationship between the organic producers and supply-chain steps. In particular, outlet markets—direct selling, wholesale, retail—have been analyzed with reference to wine categories (red, white, *novello*[17], *barrique*, aged, etc.) showing volumes and prices.

Next, the market distribution of sales has been analyzed, evaluating the share of the domestic—local, regional and national—and foreign market. Finally, marketing channels were investigated to figure out their importance on domestic and international markets, their weakness and their conflicts among wine-growers and other organic supply-chain steps.

The analysis of marketing channels shows that 36% of total wine is sold to wholesalers, 22% to bottling companies and 20% through direct selling to consumers or home delivery (Figure 4).

The sales distribution for market area was then studied with reference to domestic—local, regional, national—and international market.

Results, depicted in Figure 4, show that organic wine sales are extending not only to domestic market but also overseas. Only one-fourth of sales is sold inside regional boundaries while 38% is sold outside the Veneto Region and 36% is often sold to the foreign market. The latter does not appear surprising, since the foreign market has recently encouraged the production of high quality wines made from organic agriculture.

The packaging analysis showed the high share of bottled wine (66%) while the wine from the cask (20%) or sold in demijohn (14%) is low. The latter packaging are significant only in regular white, red or local wines.

Next, the survey analysis has focused on sale price according to market channel, area and packaging per type of wine.

The distribution for market channel shows a significant price increase in all type of wine, excluding novel and sparkling wines. The price catches values of approximately 0.82–0.97 EUR/liter in red and white wines and nearly 2 EUR/liter in local wines and barrique. The price premium is relevant in catering/restaurant channel for white and local wines, and in wholesale/LSR channel for red wines and barrique.

The distribution for market area has also shown changes in price. The premium price reaches about 1 euro/liter in white wines and 1.13 EUR/liter in the red ones. The regional market plays a positive role in the premium price of white wines, while the national one influences the red and local wines. Sales of both domestic and foreign markets seem to increase the price of white wines.

Eventually, the distribution of prices for packaging has confirmed the greater value for bottled wines which premium price reaches about 1 euro/liter for white wines and about 1.4 euro/liter for red ones.

Next, the market has been explored to figure out how marketing channels are used by organic wine-makers.

---

[17] Novello wine (or *noveau*) is obtained from the intracellular fermentation of the entire grapes placed in special stainless steel vats in absence of oxygen and saturated of carbon dioxide. This process is technically called carbonic maceration. The wine has normally a fruity bouquet, a clean red colour and a fresh taste.

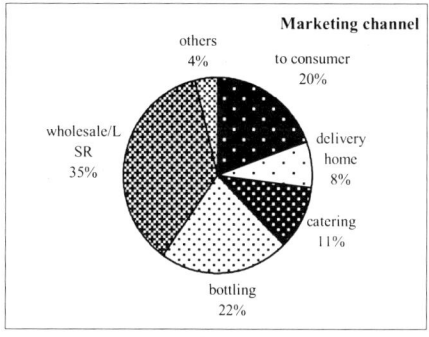 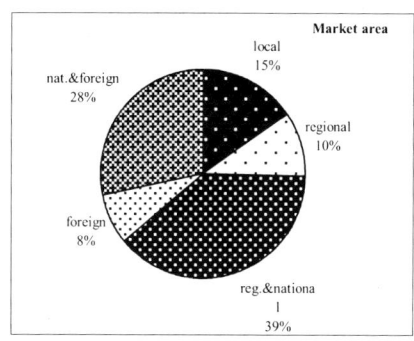

Figure 4. Sales distribution per marketing channels and market area (volumes). (Defrancesco *et al.*, 2002)

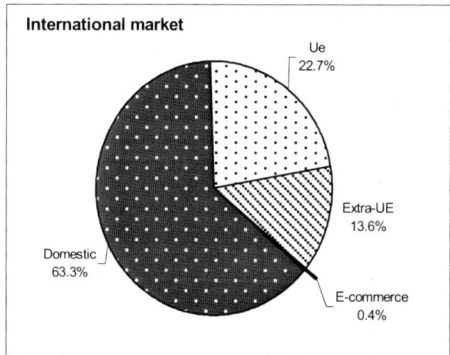 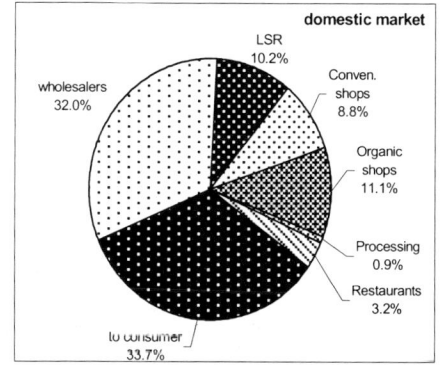

Figure 5. Distribution of organic wine production on international and domestic market. (Defrancesco *et al.*, 2002)

First, the distribution of buyers among domestic, EU, and extra-EU market has shown high employment in marketing channels such as the direct selling, wholesaler and also the LSR, while the catering/restaurant and conventional retail channels are definitively less important. Sales to traditional retail are rather turned to organic shops than conventional ones.

Almost two-thirds of organic wine is sold into the domestic market, 20% into EU-market and the remaining part is exported to extra-UE countries (Figure 5). So far, e-commerce is not an important outlet market since it involves only 0.4% of the total sales, but it is an important marketing tool utilized to improve the image of the winemaker and communicate it to new potential customers.

In the domestic market, almost one-third of organic wine is sold directly to final consumers, one-third is bought by traditional wholesalers and one-third is sold through other channels: LSR (10%), traditional retail (about 20%) restaurants (about 3%) and processors (less than 1%). Sales to traditional retailers are equally divided between conventional and organic shops.

4.6. Marketing strategies

Marketing strategies are strictly linked to reasons driving wine makers to organic farming:
1. to fill up a new market segment quickly;
2. to diversify production.

The growth in organic agriculture supported by an increasing demand and 10 years of EU grants has gradually expanded the EU organic market over its starting niche size. This result is particularly felt by companies operating close to final consumers.

The diversification of production has been carried out through product innovations, increasing product range and/or specializing in organic product. The elasticity with respect to income is higher than conventional wines. Actually, the greater product elasticity is considered by most of organic wine-makers as more important than expectations of gaining larger margins.

So far, the conventional wine market has reached maturity while the competition has increased even in high quality wines. The risk of shutting-down is quite high in the conventional wine market. Obviously, the potential risk of organic wine market is lower but its profitability is considered lower than the conventional wine market.

Once entering into organic agriculture, wine-makers have defined their own marketing strategies according to constraints coming from socio-economical environment, market and competitiveness. The decisions on marketing strategies are strongly influenced by the distribution of channels: wholesalers, LSR, catering/restaurants, direct selling to consumer.

The wholesale channel is not economically appreciated because of low profitability, long-term payments and, consequently, low cash flow. Wholesale is also considered inadequate because its logistics do not meet supply features for organic products: high volumes required by wholesalers need a supply concentration which is now infeasible (due to the small size of the organic market).

The LSR is seen as a better outlet market because it can increase the consumer loyalty and rapidly expand organic sales. The organic market growth is necessarily accomplished by the LSR, which can make investments in promoting organic products and can reach many consumers. The problem still remains that most of the mass is supplied by individual wine-makers. The wine makers should combine their production in order to increase their market power and profitability.

The catering/restaurant channel is considered a good outlet market except for long-term trade credits. The retail organic shops are also considered a good marketing channel, but they sell small volumes and do not meet fast-growing market equirements.

Direct selling to consumers is increasing but poor logistic scannot be overcome due to the limited financial resource of individual wine-makers.

The marketing strategies employed by organic wine companies are reported in Table 4. Main marketing strategies are price, product range and promotion in order to consolidate and strengthen the market share, while only a half of holdings use advertising.

Table 4. Marketing strategies (in %)

	Type of holding			
	only grapes	grapes/wine	only wine	marketing
Marketing strategies*:				
– price		71.4	100.0	100.0
– range		64.3	100.0	100.0
– marketing channel		42.9	100.0	100.0
– advertising		14.3	50.0	50.0
Why it has chosen these strategies?*				
– to consolidate the home market		18.2	50.0	100.0
– to consolidate the foreign market		54.5	50.0	50.0
– to expand the foreign market		72.7	100.0	50.0
Stability of the price				
– no	66.7			
– yes	33.3	100.0	100.0	100.0

* The percentages do not add to 100 because there are multiple responses.
Source: Defrancesco et al., 2002.

Wine-maker holdings, selling most of their production directly to final consumers and restaurants, do not apply specific marketing strategies, excluding product range and price policy. All holdings are generally opened to foreign markets, because of favorable perspectives in expanding exports.

The organic wine holdings can be seen as a "nearly" price-taker, since the price is strongly influenced by the buyer and volume exchanged and, more generally, by competitor prices. Consequently, the price is only a "defensive strategy" covering full production costs. This price weakness leads to a volatility especially on grape price paid by upper stream holdings. Only consumer-oriented holdings seem able to apply a mark-up on full production cost.

The marketing strategies for widening the product range, i.e. by introducing product innovations, are mostly aimed to reduce the market and production uncertainty. In both direct selling and the LSR channel, wine-makers widen the product range to avoid having their customers contact competitors. In medium and/or small size holdings a wide range is considered a strategy reducing marketing and production risk. In big size holdings, a greater production mix increases price and improved market power by reducing the average cost of production.

The communication strategy analysis shows a greater interest in the business brand instead of the collective one. A firm's own brand is considered more effective in increasing consumer loyalty and it is appreciated as another quality certification.

Organic certification is also perceived as an effective communication strategy to final consumer, even if the LSR has recently introduced their organic private labels certified by its own and control system (ISMEA, 2001; ISMEA, 2002).

Many firms use traditional advertising (catalogue, etc.) and Web pages to introduce their business to final customers. Over 56% of firms participate in exhibitions, especially those specialized in wine and/or the organic products. This strategy is considered effective in promoting sales on foreign markets because it improves the wine-maker image, especially in firms close to the final consumer.

Strategies for improving distribution concern, first, the reduction of supply fragmentation and, second, the diversification of marketing channels. Eventually, such strategies are aimed to improve both the market power and profitability of wine-makers, while volume and product range meet customer requirements. The diversification of the marketing channels is aimed to reduce marketing risks because of the increasing competitiveness of holdings entering this market.

Strategies for strengthening market share are focused on product quality and business brand, whereas, almost one-fourth of holdings concentrate their management on cost reduction and product innovation strategies.

5. CONCLUDING REMARKS

After ten-years of application of the EU regulation 2092/91, the Italian market of organic products is growing at a fast rate as a result of institutional and farmer actions. Recently, the market has been boosted by fast increasing organic consumption.

The overall wine sector, after the methyl alcohol crisis during 1980s, has been realigned towards high quality productions.

So far, the Italian organic wine production has been considered a market niche. Actually, the organic wine in Veneto Region is still limited in respect to the total wine market, but it is growing and looking for a right consumer's appreciation. Still, the growth path cannot be easily forecast.

On the supply side, the production is rather fragmented, and it does not meet the demand not only as volume supplied but also as logistics, distribution and marketing requirements. In particular, organic wine-makers should take advantage of this market segment which is gradually losing its niche size because of emerging consumption.

This survey has showed some aspects about organic wine-growers and about organic wine supply in the Veneto Region.

On the farm side we summarized the following points:

- the size of organic wine farms are greater than the average regional one; the technology employed and innovation adopted are sophisticated because of high managerial skills and reorganization of vineyards and wine-making process;
- on the economic side, cultivating costs and wine-making costs are respectively greater and equal to conventional farming. There is also an increasing cost instability among organic farms because of different technologies;
- information on technical and legal issues are partially satisfied by regional institutions. There is still a lack of information for making commercial and marketing strategies.

On the supply side and marketing channels it is observed that:

- the supply is strongly affected by climatic conditions and market performance. Often price volatility and/or yield variability reduce organic wine profitability: small wine-growers give up the wine-making process and sell grapes directly (in the short run), while big ones increase high quality vineyard (DOC wines);
- the wine-makers are able to capture a premium price by differentiating their sales among customers, area and packaging.

- the LSR seems to be an appropriate outlet market for organic productions because it can increase the consumer loyalty. It can also make investments in promotion and can further expand organic consumption;
- catering/restaurants and organic shops are the traditional outlet markets but their volume sold does not meet fast-growing market requirements;
- the direct selling to the consumers is limited by poor logistics, which cannot be improved by single enterprises because of high investments and small demand;
- the traditional wholesale channel is not affected by logistic weakness. However, the supply fragmentation increases wholesalers market power reducing the organic wine price and, consequently, making it less profitable for the farmer.

All wine-maker holdings are opened to foreign markets and the analyses on marketing strategies have shown two different behaviors:

- holdings in the wine-making process and trade activities employ marketing strategies such as price policy, widening the product range, promotion on distribution in order to consolidate and/or to strengthen their own market share, while advertising is still limited because of low financial resources;
- having most of production sold directly to final consumer, holdings in farming or wine-making activities do not employ specific marketing strategies, excluding price policy and widening range.

At the supply-chain level, the survey have showed a weak market power of upper stream activities related to down stream ones both in terms of price volatility and high marketing uncertainty:

- firms using wholesale channel suffered a competitive disadvantage because the price is strongly affected by small volumes and prices fixed by competitors. These wine-makers are "price-takers";
- consumer-oriented firms use a marketing strategy based on price (price-maker), by applying a mark-up on the production cost and differentiating it according to customer type and individual demand;
- the reduction in marketing uncertainty is pursued through two strategies: 1) by a diversification of marketing channels avoiding new producers entering to this market; and 2) by widening the product range avoiding own customers contacting competitors.

As far as the communication strategy, it is observed that:

- the business brand is better than the collective one, because it can increase consumer loyalty, it can be also be appreciated by consumers as an intrinsic quality cue;
- the diversification in organic wine production has been accomplished by a quality differentiation shifting from table to high quality wines (DOC);
- the organic certification is perceived as an effective communication strategy to the final consumer. There is still a lack in consumer information on production processes and organic product features. This difficulty may be outweighed by promotional and information campaigns, financed by EU, national or regional institutions;

- the traditional promotional strategies (specialized exhibitions on wine or organic products) are widely spread among wine-makers, especially at the farm level, even if innovating and less-expensive strategies such as Web pages or e-commerce can be successfully employed.

Usually, marketing strategies are focused on sales and customers are trying to strengthen quality standards and the business brand. The farm management is oriented to cost control and wine-making innovations (process and product).

The analysis of the market development path provides additional information. In the 1980s, pioneering organic farmers had been slowly followed by other wine-growers. During the 1990s, the introduction stage has been followed by the growth stage, while the organic farmer specialized in organic grape production. At the same time, distribution channels have changed according to production and consumption growth, i.e., as the production increased, the outlet market changed from direct selling to organic stores and extended from agricultural to urban areas. Eventually, the fast increasing production has also reached many supermarket chains extending to regional and international market. This path has been driven mainly by EU rules and financial supports, and more recently by the rapidly increasing demand.

Wine-makers marketing to LSR have to cope with the increasing competitiveness. The wine-maker behavior may be explained by several competitive models (Boatto, 2002):

1) the organic wine production is considered a *strategic business activity*. For example, leader wine companies adopt this marketing strategy for differentiating product rather than increasing their profitability;
2) old organic wine-growers (pioneers and starting ones) face competitors by improving their production efficiency (technology), product quality and marketing to LSR as well (*competitive strategy*);
3) wine-growers specialize in organic wine production fitting their market segment and focusing especially on technology (*strategic business unit*).

However, the development of this market should be accomplished by institutional supports (rules and grants) aiming to improve the market transparency and competitiveness conditions. For example, the consumer awareness on differences between organic and conventional product should be encouraged (the attention paid by consumers to organic products is proportional to the information available). The institutional actions should also ensure transparent market rules avoiding confusion between production logos (organic, ecological, natural, etc.) and collective brands (DOC, IGT[18]).

In this market stage the role of the wine-maker is important *"per se"*. Due to high innovation standard, the growth path of organic wine-makers depends strictly on their strategic ability and do not fall into canonical competitive models.

---

[18] In Italy the IGT "Indicazione geografica tipica" or "Typical Geographic Indication" is a collective label referring to table wines which production area is wide and production process is less restrictive than DOC wines. The IGT label is similar to the French term "Vin de Pays" or the German term "Landwein".

6. REFERENCES

AIAB, 2004, *Proposta di disciplinare per il vino biologico*, online document; http://www.aiab.it.
Boatto, V., 2002, Da mercato di nicchia a mercato generale, in: *Il mercato della carne e del vino da agricoltura biologica nel Veneto*, V. Boatto and G. Favaretti, eds., Veneto Agricoltura, Regione Veneto, Venezia, pp. 10–18.
Castaldi, R., 2000, Il vino proveniente da coltivazione biologica, *L'informatore Agrario*, **28**: 59–61.
Compagnoni, A., 2001, *Coping with growth*, paper presented at the "Organic Food and Farming: towards partnership and action in Europe" Conference held in Copenaghen 10–11 May.
Defrancesco, E., 2003, I cambiamenti dell'agricoltura veneta visti attraverso i dati censuari. Il sistema delle produzioni di qualità: verso un sistema integrato regionale?, in: *Rapporto 2003 sul sistema agroalimentare del Veneto*, Veneto Agricoltura, Regione Veneto, Venezia, pp. 275–402.
Defrancesco, E., Galletto, L., and Rossetto, L., 2002, Le Imprese Biologiche Viticole nel Veneto: Caratteristiche Strutturali e Strategie Commerciali, in: *Il mercato della carne e del vino da agricoltura biologica nel Veneto*, V. Boatto and G. Favaretti, eds., Veneto Agricoltura, Regione Veneto, Venezia, pp. 128–170.
Didero, L., 2000, Biologico verso la vetta, *Largo consumo*, **12**: 22–29.
FAO/WHO, 2001, *Codex Alimentarius Commission*, 24th session, Geneva, July; http://www.codexalimentarius.net.
Foreign Agricultural Service (FAS), 2004, online data; http://www.fas.usda.gov.
Gallas, P., 2000, Un mercato in costruzione, *Largo consumo*, **2**: 22–34.
Galletto, L., Favaretti, G., Gennari, A., and Chung-Pei, Y., 2002, La situazione dei prodotti dell'agricoltura biologica con particolare riferimento alle carni ed al vino, in: *Il mercato della carne e del vino da agricoltura biologica nel Veneto*, V. Boatto and G. Favaretti, eds., Veneto Agricoltura, Regione Veneto, Venezia, pp. 19–106.
Gardner, B., 2000, *Organic Food in the European Union: Production, Consumption and the Development of Markets*, Agra Europe, London; http://www.agra-europe.it.
Greene, C.R., 2001, US Organic farming emerges in the 1990s: adoption of certified systems, USDA-ERS, *Agriculture Information Bulletin*, n. 770.
Haccius, M., and Lünzer, I., 2001, *Organic agriculture in Germany in 2001*, Stiftung Ökologie & Landbau (SÖL), Bad Dürkheim, Germany; http://www.organic-europe.net.
Häring, A.M., Dabbert, S., Aurbacher, J., Bichler, B., Eichert, C., Gambelli, D., Lampkin, N., Offermann, F., Olmos, S., Tuson, J., and Zanoli, R., 2004, Organic farming and measures of European agricultural policy, in: *Organic farming in Europe: Economics and Policy*, Volume 11, Stuttgart-Hohenheim.
IFOAM, 2000, *Basic Standards for Organic Production and Processing*, Ifoam General Assembly, Basel, Switzerland; http://www.ifoam.org.
ITC-International Trade Center UNCTAD/WTO, 2001–2004, online data, http://www.intracen.org/home.htm.
ISMEA, 2001, *La spesa alimentare di prodotti biologici*, Rome, online document, http://www.sinab.it
ISMEA, 2004, *Lo scenario economico dell'agricoltura biologica*, Rome.
ISMEA, 2002, *La spesa per i prodotti biologici confezionati: 2002*, Osservatorio consumi, Rome; http//:www.ismea.it.
Kopfer, P., and Willer, H., 2001, *Organic Viticulture in Germany*, presented at the "Biobacchus, International Organic Wine Conference" held in Frascati 5–6 May.
Lunati, F., 2002, *Il biologico in cifre 2002*, Rapporti Biobank, Distilleria EcoEditoria, Forlì.
Lunati, F., 2004, *Il biologico in cifre 2004*, Rapporti Biobank, Distilleria EcoEditoria, Forlì.
Mingozzi, A., and Bertino, M.R., 2005, *Tutto Bio 2005*, Distilleria EcoEditoria, Forlì.
Pinton, R., 2001, Dove va il vino?, *Bioagricultura*, **71**: 33–37.
Santucci, F.M., and Pignataro, F., 2002, *Organic farming in Italy*, paper presented at the OECD workshop on organic agriculture held in Washington D.C., 23–26 September.
SENASA, 2004, Situación de la Producción Orgánica en la Argentina año 2002, Buenos Aires; http://www.senasa.ar.
Sgarbi, S., 2001, Introduzione, in: *Guida ai vini biologici d'Italia 2001*, Attorre, A., ed., Tecniche Nuove, Rome.
USDA Agricultural Marketing Service, 2000, *National Organic Program: Final Rule*, Washinghton DC, USA; http://www.ams.usda.gov/nop.
USDA-ERS, 2002, *Harmony between agriculture and the environment: current issues*; http://www.ers.usda.gov.
Yussefi, M., and Willer, H., 2004, *The World of organic agriculture: statistics and emerging trends*, Stiftung Ökologie & Landbau (SÖL), Bad Dürkheim, Germany; http://www.soel.de.

THE CONSUMERS' PERSPECTIVE

INVESTIGATING PREFERENCES FOR ENVIRONMENT FRIENDLY PRODUCTION PRACTICES
Taste segments for organic and integrated crop management potatoes in Italian households

Riccardo Scarpa, Fiorenza Spalatro, and Maurizio Canavari[*]

SUMMARY

This paper reports some preliminary results on a mixed logit random utility analysis of conjoint data from consumers' preferences for agricultural products. The data were collected via a telematic sample representative of Italian households. The survey instrument was implemented via a computer supported system. A multivariate normal full correlation structure is imposed in the mixed logit estimation and the implications of such a taste structure are examined.

1. INTRODUCTION

Environment friendly production methods represent a way to meet society's need for a lower-impact agriculture and a way to cater to a category of consumers with particular preferences. Among environment friendly products, two categories appear of special interest: organic food and food produced using integrated crop management (ICM). The latter may be seen by the consumer as a "softer" alternative to the former, as it allows the

---

[*] Riccardo Scarpa, Department of Economics, Waikato Management School, The University of Waikato, New Zealand, corresponding author: rscarpa@mngt.waikato.ac.nz, Fiorenza Spalatro, Economics Department, University of Siena, Italy, Maurizio Canavari, Department of Agricultural Economics and Engineering, *Alma Mater Studiorum*-University of Bologna, Italy. The paper is the product of joint work by the authors. F. Spalatro wrote section 1, M. Canavari wrote sections 3 and 5 and R. Scarpa wrote sections 2 and 4. This research paper was presented at the 8th Padova-Minnesota Joint Conference on Food, Agriculture, and the Environment, August 25–28, 2002, Red Cedar Lake, Wisconsin, USA. The authors wish to acknowledge the various comments of the participants, but remain solely responsible for the remaining shortcomings.

controlled use of pesticides and relies on much less stringent production restrictions and protocols. Finally, quality certification programs which are now available to certify specific production practices have been applied to food produced with both the above methods.

While the determinants of price mark-ups for organic products are quite well understood (La Via and Nucifora, 2002), despite much empirical work in Italy (Antonelli, 1996; Canavari et al., 2002; Chinnici et al., 2002; Cicia and Perla, 2000; Mora Zanetti, 1998; Santucci et al., 1999; Gregori and Prestamburgo, 1996), the structure of household preferences for environment-friendly production methods is still poorly understood. A number of issues make such an understanding complicated. Some are related to the quality of data available from conventional sources of market transactions and some due to the perception that households have of certification methods for agricultural products. At least in Italy, revealed preference data are particularly difficult to collect without the co-operation of large organized retailers, who tend to shun away from collaboration with university researchers on the grounds of their right to protect their marketing strategies from indiscrete eyes. So, revealed preference data, of the kind used, for example, by Bonnet and Simioni (2001) or Jain et al. (1994), are either very difficult or extremely expensive to obtain.

An alternative to revealed preference data is represented by stated-preference data, which are often considered more informative because of the higher flexibility that the experimental design can provide (Louviere et al., 2000). Such an approach is commonly employed for studying preferences for fruit and vegetable products (van der Pol and Ryan, 1996).

Following seminal studies in the United States of America (Misra et al., 1991), recent research based on stated preferences have investigated, in some detail, the structure of preferences for some relatively high-value products from Italian agriculture, such as strawberries and table grapes (Scarpa and Spalatro, 2001) and extra-virgin olive oil (Cicia et al., 2002; Del Giudice and Scarpa, 2002; Scarpa and Del Giudice 2004). One common finding of these studies is that conventionally available socio-economic covariates go only a small way towards explaining taste-heterogeneity. A larger fraction of the variation in taste-intensity remains "unobserved", yet it can be accounted for "unconditionally" on measurable socio-economics covariates by means of random parameter models. How much of these findings extend from high value-added products, such as fruit, cheese and olive oil, to low value products, such as potatoes and other starchy products, remains an empirical question. As the market for products obtained with environment friendly methods grows, the answer to such a question is of increasing interest to policy makers, who seek to find alternative ways to boost farmers income and decrease the negative impacts of conventional production methods which rely on higher levels of chemical input use.

Some basic questions pertain to a) the way taste heterogeneity defines market shares, b) what is the price differential that different products are likely to command in the market place, and c) what kind of modeling approach performs best in describing taste variation.

The present paper reports on an investigation of preferences for potatoes based on stated preference data collected from a representative sample of 2,000 Italian households using the telematic network administered by AC-Nielsen S.p.A market research unit. The choices were derived by asking consumers to rate preferences for pairs of alternatives.

Each alternative was described on the basis of standard commercial attributes as well as environment friendly production methods, such as organic and ICM. Quality certification programs and area of production (domestic/foreign) were also included to investigate the intensity of home-bias which appeared to be significant in other studies (Scarpa et al., 2001, 2005).

Preliminary results indicate that the observed data support the presence of unobserved more than conditional heterogeneity. Second, there is evidence that the distributions of taste intensity for organic, ICM and quality certification (QC) share a common correlation structure, rather than being independent of each other. Third, the largest price differential (2.47 EUR/kg) is commanded by organic potato production in the relatively small (10%) share of the market of those who favor organic and ICM potatoes, but dislike QC. Finally, the mean willingness to pay (WTP) for ICM potatoes is always inferior to that for organic ones, and is highest in the largest market share (41%) of those who favor all three production systems.

The remainder of the paper is organized as follows. In section 2 we briefly mention the theory underlying the empirical analysis. In section 3 we illustrate the process by which the stated preference data were collected. In section 4 we describe the econometric analysis and discuss the results. We report our conclusions in section 5.

2. THEORY

Mixed logit (MixL) estimation (or random parameter logit) stands as probably the most significant advance in random utility discrete choice analysis (McFadden and Train, 2000). MixL estimates have been made possible since methods for simulated maximum-likelihood have become a practical alternative to practitioners. An appealing way to interpret MixL estimates is in the context of random utility theory, which is a well understood paradigm to interpret discrete choices and need not be illustrated here (see Train 2003 for an up-to-date treatment of logit modeling and random utility theory). This advance in modeling discrete choices significantly enriched the amount of information that can be obtained from conjoint analysis surveys, especially in the context of choice ridden with preference heterogeneity, such as in food-related choices. The focus of this paper is on the additional information that can be gleaned from a full correlation structure across distributions of taste intensities vis-à-vis the more conventional way to represent taste heterogeneity in these models, i.e. through the inclusion of interaction variables with socio-economic covariates.

With few exceptions (Train, 1998; Scarpa and Spalatro, 2001; Del Giudice and Scarpa, 2002; Scarpa and Del Giudice, 2004; Scarpa et al. 2003, 2005), mixed logit models are estimated under the assumption that taste intensities are distributed independently of each other (Layton, 2000; Garrod et al., 2002; Hensher and Greene, 2001). This is often a necessity because correlated structures are often more difficult to estimate from a given dataset, and may require the researcher to impose restrictions in the covariance matrix in order to reach convergence. However, correlated taste structures are substantially informative in that they allow researchers to derive joint distribution functions and hence joint taste-segments probabilities. For markets for which taste segmentation is not well developed or understood, such as for the market of potatoes produced with environment-friendly methods, such a structure can help uncover the size of the segments and the expected WTP for the production related product attributes, hence helping the identification of those segments for which it is potentially profitable to specialize the production.

The notion that taste intensity for agricultural products varies across consumers (households) is quite intuitive and need not be argued further (Moro and Sckokai, 2000). For some specialized (local and traditional) products taste variation for origin and ways of production is shown to vary significantly across cities, possibly in relation to the degree of ethnic heterogeneity (Scarpa and Spalatro, 2001; Del Giudice and Scarpa, 2002; Scarpa and Del Giudice 2004). From the researcher's perspective however, an issue of interest is the development of an operational approach to obtain policy relevant information from such a variation. For example, in a context of policy design, it is often useful to be able to obtain descriptions of taste variation conditional on socio-economic co-variates. Market shares are then estimated on the basis of the population distribution of such covariates. However, taste variation for food products may well be orthogonal to commonly measured socio-economic co-variates. In the absence of such a relationship, the estimation of a correlation structure, and hence of a joint distribution, still identifies the market shares for various taste segments, and conditional on these, WTP for food attributes can be obtained.

The identification of segments for which it is potentially profitable to specialize the production can be achieved by combining information on the segment sizes, WTP distributions and costs of production (at the retail level) and of certification programs. Let c^* be the cost of production per unit of weight of a particular potato type identified by taste conditions θ, where θ is a $1 \times k$ vector, then the potential market share is going to be proportional to $Pr(WTP>c^*|\theta>0)$. In other words, it is given by the share of people in a taste segment who are willing to pay above the cost of production, given that their taste intensity parameter is positive (i.e. they like the attributes defining the taste segments of the product). Under independence of taste intensities, by definition this is the product of marginals:

$$Pr(WTP>c^*|\theta>0) = \Pi_{i=1}^{k} Pr(WTP>c^*|\theta_i>0), \qquad (1)$$

while with correlation across θ_i this is:

$$Pr(WTP>c^*|\theta) = \int_1 \ldots \int_k f(WTP>c^*|\theta_1,\ldots,\theta_k) \, d\theta_1,\ldots,d\theta_k \qquad (2)$$

Such a probability can be estimated from the parameter regulating the behavior of the joint distribution of taste intensities. Suppose we assume that $\theta \sim \Phi(T, \Omega)$, where Φ is the multivariate normal with mean T and variance covariance matrix Ω. Then, one relatively simple way to simulate those probabilities is using the Cholesky decomposition of $\Omega = C^T C$ and simulating a sufficiently large vector of variates $\mu + C'z$, where μ is a $1 \times k$ vector, C is the $k \times k$ lower triangular Cholesky matrix and z is an $k \times r$ vector of standard normal variates. The distribution properties of such simulated variates can then be analyzed to derive the statistics of interest.

3. DATA

The stated preference data were collected from a sample of 2,000 households representative of the Italian population of consumers. The sample and the administration of the survey were conducted by AC Nielsen Market Research Italy on behalf of a research project lead by the EU. The survey administration was conducted via PC terminals installed in the homes of respondents. The household member in charge of grocery shopping was asked to take part and each respondent was asked to provide a rating from a scale of 21 points for only one pair of product alternatives. The 11 mark in the scale identified indifference between the two alternatives. The marks 10 to 1 identified increasing preference for the first alternative that appeared to the left of the PC screen, while the marks 12 to 21 indicated increased intensity of preference for the alternative on the right of the screen. Choice sets were designed by randomly pairing alternatives from a main effect partial factorial orthogonal design. The number of factors was 12, each expressed at 2 levels, except for price that was expressed at 3 levels. For such a set-up we obtained a total number of 16 profiles using the "orthoplan" routine in SPSS.

4. ECONOMETRIC ANALYSIS AND RESULTS

In a recent paper, Hensher and Greene (2001) provide an extended review on the increased complexity that analysts must cope with when estimating mixed logit models, with particular reference to the need to avoid distributional specifications with meaningless or counter-intuitive behavioral implications. In our case we first postulated and then tested that the attributes of relevance for our investigation display heterogeneous taste-intensities across the population of Italian households. We focused on taste-heterogeneity for three attributes: *organic*, *integrated crop management* and *quality certification*, while all the other attributes were assumed to have fixed taste-parameters. Taste for price (or loss of income) was considered fixed[1]. Table 1 reports the 6 models estimated by (simulated) maximum likelihood with 120 Halton draws (Train, 1999). The models are multinomial logit (MNL), mixed logit without correlation (MixL), and mixed logit with correlation (MXLC). The last three columns report models with 20 interaction variables between potato attributes (Price, ORG, ICM and QC) and socio-economic (_SE) covariates (sex, age and education level of the respondent and income and size of the household). For the sake of brevity, we only report the attribute estimates (detailed estimates can be obtained from the corresponding author). In these models, age of the respondent was negative and significant in its interaction with QC, suggesting that older respondents appreciate quality certification less than younger ones. It was also positively significant in its interaction with ICM, and nearly so in its interaction with ORG, suggesting an

---

[1] The distributional assumptions concerning the price parameter are problematic because they determine the distribution of WTP for each parameter, that, in commonly employed linear utility models, is given by the negative of the ratio of the parameter of interest and the parameter for price. For some choices of distribution the distribution of the ratio has infinite first central moments. For example, when both are normally distributed. We actually tested for the log-normal (suggested by Train) and the constrained triangular (suggested by Hensher and Greene) distributions for the price parameter, but departing from other findings in other empirical settings (Scarpa *et al.*, 2003; Train, 1998) in both cases we found the estimates for the spread parameters to be insignificant, hence rejecting the null hypothesis of heterogeneity for price.

Table 1. Simulated Maximum-likelihood estimates of logit models

Variables	MNL	MixL	MXLC	MNL_SE	MixL_SE	MXLC_SE
Log-L	−987.15	−981.79	−978.33	−975.67	−970.35	−967.54
Adj. Pseudo-R squared	0.114	0.117	0.118	0.113	0.116	0.117
Organic μ	0.58 (0.01)	0.97 (0.25)	1.13 (0.27)	−1.11 (0.68)	−1.29 (1.03)	−1.69 (1.32)
Organic σ	----	2.32 (0.75)	3.21 (0.93)	----	2.10 (0.62)	3.15 (0.90)
Integ. Crop Manag. μ	0.21 (0.09)	0.34 (0.13)	0.36 (0.17)	−1.90 (0.69)	−2.16 (0.83)	−2.72 (1.15)
Integ. Crop Manag. σ	----	0.67 (0.76)	1.02 (0.75)	----	0.13 (1.18)	0.89 (0.76)
Quality Certification μ	0.56 (0.08)	0.77 (0.19)	0.87 (0.19)	1.17 (0.56)	1.67 (0.76)	2.06 (0.95)
Quality Certification σ	----	0.84 (0.77)	0.10 (4.75)	----	0.47 (1.13)	0.07 (5.11)
White Pulp	−0.33 (0.08)	−0.41 (0.12)	−0.48 (0.13)	−0.34 (0.08)	−0.40 (0.19)	−0.49 (0.13)
Large Size	0.19 (0.10)	0.23 (0.12)	0.28 (0.14)	0.23 (0.09)	0.26 (0.18)	0.32 (0.14)
Medium Size	0.26 (0.09)	0.34 (0.13)	0.40 (0.14)	0.27 (0.09)	0.33 (0.12)	0.39 (0.14)
Poor Appearance	−0.60 (0.08)	−0.78 (0.14)	−0.88 (0.14)	−0.62 (0.08)	−0.75 (0.13)	−0.88 (0.14)
5 kg bag	0.15 (0.08)	0.24 (0.11)	0.30 (0.13)	0.17 (0.08)	0.24 (0.11)	0.31 (0.13)
Domestic origin	0.84 (0.08)	1.14 (0.20)	1.29 (0.18)	0.84 (0.09)	1.07 (0.19)	1.26 (0.18)
Price	−0.70 (0.23)	−0.97 (0.33)	−1.05 (0.35)	−0.45 (1.68)	−0.35 (2.07)	−0.29 (2.52)

Source: authors' calculations on survey data. Standard errors in brackets.

effect in the other direction for integrated *crop* management and organic production of potatoes. The last effect to be found significant was the interaction between size of the household and ORG, suggesting that respondents with larger households appreciate organic production relatively more than others. This is perhaps justified if parents with many children try to safeguard their children's health by buying organic rather than conventional.

Taste heterogeneity was first assumed to be independent across attributes, and the significance of standard deviation estimates was assessed by computing the likelihood ratio test where the restricted model is the MNL and the unrestricted is the mixed logit without correlation (MixL). Then a full correlation pattern was allowed to test for jointness in taste distributions (MXLC). The significance of the off diagonal elements of the Cholesky matrix was assessed by computing the likelihood ratio test where the restricted model is the mixed logit without correlation (MixL). Table 2 reports the various likelihood ratio test statistics.

A number of points can be made by looking at the values in the table. First, it is clear that the data cannot reject the null hypothesis of unobserved heterogeneity, and in particular its joint form, since the models MXLC and MXLC_SE achieve significantly

Table 2. Chi-square statistics for the various restrictions

		\multicolumn{6}{c}{Restricted Models}
		MNL	MixL	MXLC	MNL_SE	MixL_SE	MXLC_SE
	Log-lik.	−987.15	−981.79	−978.32	−975.67	−970.35	−967.54
Unrestricted Models	MixL	0.013	----				
	MXLC	0.007	**0.074**	----			
	MNL_SE	*0.291*	----	----	----		
	MixL_SE	0.071	*0.295*	----	**0.014**	----	
	MXLC_SE	0.046	*0.197*	*0.365*	0.012	*0.132*	----

Source: authors' calculations on survey data

higher log-likelihood values (statistics in bold) than their immediate counterparts MixL and MixL_SE. Secondly, the addition of socio-economic covariates via the 20 interaction variables never represents a significant improvement on their immediate counter-parts (statistics in underlined italics) despite involving substantial over-parameterization. Accounting for heterogeneity in an unobserved fashion provides a much better fit and a much smaller addition of parameter estimates than accounting for it in a form conditional on socio-economic covariates, as can be seen comparing the log-likelihood statistical improvements obtained from the basic MNL to the MixL (or MXLC) to that obtained from MNL to MNL_SE. This result is common to other studies (e.g. Scarpa and Spalatro, 2001).

Having established that the null hypothesis of joint taste parameter distribution is supported in the data, we concentrated on the model without socio-economic variables (MXLC) and used the parameter estimates of the joint distribution of tastes to simulate taste segment shares, and marginal and conditional distributions of WTP for each of the three production related attributes. These are reported in Table 3.

The largest taste-based market share is about 40% and includes those who enjoy ORG, ICM and QC in potatoes in segment A. In this large segment the mean WTP for ORG is 1.42 EUR/kg. ICM, the soft alternative to ORG, has a mean WTP of 0.80

Table 3. Estimated taste-based market shares and mean WTP for attributes by segment

Segment	β > 0	β < 0	Shares	E(WTP|β) in €/kg Organic	ICM	QC
A	Organic, ICM, QC		0.4086	1.42	0.80	0.64
B	QC	Organic, ICM	0.2257	-------	-------	0.60
C	Organic, ICM	QC	0.0981	2.47	0.52	-------
D	Organic	ICM, QC	0.0941	0.87	-------	-------
E	ICM, QC	Organic	0.0806	-------	0.31	1.22
F		Organic, ICM, QC	0.0563	-------	-------	-------
G	Organic, QC	ICM	0.0366	0.30	-------	0.19
H	ICM	Organic, QC	0.0000	-------	-------	-------

Source: authors' calculations on survey data

EUR/kg. QC, a potentially additional attribute to either ORG or ICM, shows a value of 0.64 EUR/kg.

The second largest segment, segment B, includes those who only like quality certification, but are not attracted to organic and integrated crop management. This segment has a mean WTP for QC approximately equal to that of the previous larger segment. The estimated size of this segment indicates that QC can potentially be supported as a marketing strategy decoupled from ORG and ICM. That is, quality certification programs which are not associated to environment friendly production methods may still win a premium price of similar magnitude for this attribute.

Three more shares of around 8–9% each are also identified. The first, segment C, includes those who are attracted by organic and ICM potatoes, but not by QC. The mean WTP that this segment shows for ORG is more than 1 EUR/kg higher than that of the largest segment. The mean WTP for ICM is about 0.30 EUR/kg lower, suggesting that respondents in this segment perceive ORG as much more desirable than its alternative ICM. The second, segment D, includes those who only like ORG, but dislike the other two. The mean WTP for ORG is considerably lower than in the other two segments, only 0.87 EUR/kg. The third share of 8% of the market, segment E, includes those who have somewhat unusual preferences: they like ICM and QC potatoes but not organic ones. These people do not seem to appreciate that the two methods are, at least in part, substitutes and are willing to pay a substantial amount for QC (1.22 EUR/kg), but relatively little for ICM (0.31 EUR/kg). A segment with similarly unusual preference is segment G, which accounts for only 3.66% and enjoys ORG and QC but not ICM, displaying only a low WTP. The segment in which none of these production attributes is appreciated (F) has an estimate share of 5.63%, while that including those disliking organic production and QC, but liking ICM shows an estimated null share.

Among the other attributes that respondents found desirable, but were treated as fixed in our analysis, it is interesting to note the large weight assigned to *domestic origin* of the product, which indicates the presence of home-bias in this market. This is similar to what has been found in other studies on different agricultural products (Scarpa *et al.*, 2001; Del Giudice and Scarpa, 2002; Scarpa and Del Giudice, 2004).

5. CONCLUSIONS

Certification of production protocols has been put forward as a means to differentiate agricultural markets and cater to differentiated market segments, while at the same time extracting more surplus from consumers and boosting farmers' income. In this context, environment friendly production methods and quality certification programs are amongst the most promising policy tools to achieve double dividends for society by diminishing externalities and increasing consumers' satisfaction and increasing demand for safe high quality food. However, the perceptions of and hence the preferences for these forms of production are still poorly investigated, especially in low value added products such as potatoes. This study presents some preliminary results from a nationwide survey in which pair-wise rating of product profiles were obtained from the member of the household in charge of grocery shopping. The sample was of 2,000 households representative of the Italian population, and the survey instrument was developed to be administered via PC.

In the analysis of the responses, we employed a random utility paradigm implemented via mixed logit and ascertain that the sample displays *joint* taste intensity distribution for environment friendly production methods such as organic, integrated crop management and quality certification. The estimated correlation structure was then employed to estimate taste-based market share and mean WTP for each desirable attribute in each share.

Perhaps the most interesting results of this analysis are those concerning the shares of some segments and within each segment the strength of the taste for some of the attributes. The results seem to indicate that only 23% of the households in Italy is not interested in environment friendly production practices, and only little more than 36% is not interested in organic potatoes. Integrated crop management, which can be seen as a "soft" alternative to organic is disregarded by 41% of the households. Quality certification is disregarded by 25% of the households and can command the same mean WTP in segments that appreciate and do not appreciate environment friendly production, hence indicating that these practices can be easily decoupled from each other.

The "core" of households that enjoy only environment friendly production methods in potatoes is made up by segment C, which represents about 9% of the households. In this segment the mean WTP for organic is nearly 5 times that for ICM, showing that perhaps there is some return from educating consumers more about the role of ICM as a substitute to organic production, perhaps through targeted advertising.

6. REFERENCES

Antonelli, G., 1996, Il mercato dei prodotti dell'agricoltura biologica: Un'indagine in un'ottica di marketing, *Rivista di Economia Agroalimentare*, **1**(1): 107–146.

Bonnet, C., and Simioni, M., 2001, Assessing Consumer Response to Protected Designation of Origin Labelling: A Mixed Multinomial Logit Approach, *European Review of Agricultural Economics*, **28**(4): 433–449.

Canavari, M., Bazzani, G.M., Spadoni, R., and Regazzi, D., 2002, Food safety and organic fruit demand in Italy: a survey, *British Food Journal*, **104**(3/4/5): 220–232.

Chinnici, G., D'Amico, M., and Pecorino B., 2002, A multivariate statistical analysis on the consumers of organic products, *British Food Journal*, **104**(3/4/5): 187–199.

Cicia, G., and Perla, C., 2000, La percezione della qualità nei consumatori di prodotti biologici: uno studio sull'olio extra-vergine di oliva tramite conjoint analysis, in: De Stefano F. (ed.), *Qualità e valorizzazione nel mercato dei prodotti agroalimentari tipici*, Edizioni Scientifiche Italiane, Napoli, pp. 237–252.

Cicia, G., Del Giudice, T., and Scarpa, R., 2002, Consumers' perception of quality in organic food: a random utility model under preference heterogeneity and choice correlation from rank-orderings, *British Food Journal*, **104**(3/4/5): 200–213.

Del Giudice, T., and Scarpa, R., 2002, *Preference-based consumer segmentation via Mixed Logit: the case of extra-virgin olive oil in 3 urban markets in Italy*, ISSEI conference, Aberystwyth, Wales, U.K., July 24, 2002.

Garrod, G.D., Scarpa, R., and Willis, K.G., 2002, Estimating the Benefits of Traffic Calming on Through Routes: A Choice Experiment Approach, *Journal of Transport Economics and Policy*, **36**(2): 211–231.

Gregori, M., and Prestamburgo, M., 1996, *Produzioni biologiche e adattamenti d'impresa*, Franco Angeli, Milano.

Hensher, D.A., and Greene, W.H., 2001, The Mixed Logit Model: The State of Practice and Warnings for the Unwary, Working Paper, School of Business, The University of Sydney. Published in *Transportation Research B*, 2003.

Jain, D.C., Vilcassim, N.J., and Chintagunta, P.K., 1994, A Random-Coefficients Logit Brand-Choice Model Applied to Panel Data, *Journal of Business and Economic Statistics*, **12**(3): 317–328.

La Via, G., and Nucifora, A.M.D., 2002, The determinants of the price mark-up for organic fruit and vegetable products in the European Union, *British Food Journal*, **104**(3/4/5): 319–336.

Layton, D.F., 2000, Random Coefficient Models for Stated Preference Surveys, *Journal of Environmental Economics and Management*, **40**(1): 21–36.

Louviere, J.J., Henshser, D.A., and Swait, J.D., 2000, *Stated Choice Methods*, Cambridge University Press, Cambridge, U.K.

McFadden, D., and Train, K.E., 2000, Mixed MNL models for discrete response, *Journal of Applied Econometrics*, **15**(5): 447–470.

Misra, S.K., Huang, C.L., and Ott, S.L., 1991, Consumer Willingness to Pay for Pesticide-Free Fresh Produce, *Western Journal of Agricultural Economics*, **16**(dec.): 218–227.

Mora Zanetti, C., 1998, La disponibilità a pagare dei consumatori per prodotti alimentari sicuri, *La Questione Agraria*, **72**: 133–163.

Moro, D., and Sckokai, P., 2000, Heterogeneous Preferences in Household Food Consumption in Italy, *European Review of Agricultural Economics*, **27**(3):305–323.

Santucci, F.M., Marino, D., Schifani, G., and Zanoli, R., 1999, The marketing of organic food in Italy, *Medit*, **4**: 8–14.

Scarpa, R., and Del Giudice, T., 2004, Preference-based segmentation via mixed-logit: the case of extra virgin olive oil across urban markets, *Journal of Agricultural and Food Industrial Organization*, volume 2, paper 7; http://www.bepress.com/jafio/vol2/iss1/art7.

Scarpa, R., and Spalatro, F., 2001, Eterogeneità nelle preferenze al consumo: il caso del biologico e della lotta integrata nell'uva da tavola e nelle fragole, *Rivista di Economia Agraria*, **56**(3): 417–450.

Scarpa, R., Kristjanson, P., Drucker, A., Radeny, M., Ruto, E.S.K., and Rege, J.E.O., 2003, Valuing indigenous cattle breeds in Kenya: An empirical comparison of stated and revealed preference value estimates, *Ecological Economics*, **45**(3): 409–426.

Scarpa, R., Philippidis, G., and Spalatro, F., 2001, *"Product-Country Images" and Preference Heterogeneity for Mediterranean Food Products: A Random Utility Analysis*, Paper presented at the 75th Conference of the Agricultural Economics Society, September 12, 2001.

Scarpa, R., Philippidis, G., and Spalatro, F., 2005, Product-Country Images and Preference Heterogeneity for Mediterranean Food Products: A Discrete Choice Framework, *Agribusiness*, **21**(3): 329–349.

Train, K.E., 1998, Recreation Demand Models with taste differences over people, *Land economics*, **74**(2): 230–239.

Train, K.E., 1999, *Halton Sequences for Mixed Logit*, Dept. of Economics, University of California, Berkeley.

Train, K.E., 2003, *Discrete choice methods with simulation*, Cambridge University Press, Cambridge, U.K.

van der Pol, M., and Ryan, M., 1996, Using conjoint analysis to establish consumer preferences for fruit and vegetables, *British Food Journal*, **98**(8): 5–12.

POTENTIAL DEMAND FOR ORGANIC MARINE FISH IN ITALY[*]

Edi Defrancesco[**]

SUMMARY

Italian demand for organic products is rapidly increasing, yet there is currently no supply of certified organic marine-fish. Moreover, in recent years marine fish farm profitability has been reduced because of competition from imported products. A pilot project was carried out in order to: a) define standards for organic marine fish farming; b) evaluate production costs in four farms, experimenting semi-extensive organic fish farming under proposed standards (seabream, *Sparus aurata* and seabass, *Dicentrarchus labrax*); c) estimate the potential demand for certified organic marine fish and consumer willingness to pay in order to figure out the profitability of a product differentiation strategy. This paper shows the economic results for production costs at the farm level and potential demand. The latter has been estimated using survey-data of 6,877 consumers by means of a questionnaire-interview carried out during an experimental organic marine fish promotion sale. Results show that organic marine fish farming could be a good market opportunity for some Italian fish farmers by improving consumer information on organic products, adopting a supply concentration strategy at the farm level and carefully managing semi-extensive-farming set up by proposed regulations.

1. INTRODUCTION

After an increase over the few past decades, Italian demand for fisheries and aquaculture products has been relatively stable since 1995[1], but is lower than in other EC

---

[*] Research supported by Uniprom under EC FIFG Financial Instrument for Fisheries Guidance (EC Reg. 2080/93), project coordinator Stefano Cataudella, University of Rome "Tor Vergata".
[**] Dept. TeSAF, University of Padova, Agripolis Via Romea, 35020 LEGNARO (PD) Italy; e-mail: edi.defrancesco@unipd.it.
[1] A temporary increase in fish consumption was observed mainly in November 2000 (+25.9%) and, declining, up to the end of the first semester 2001 as a substitution to meat consumption due to Bovine Spongiform Encephalopathy (BSE) and other terrestrial animal diseases.

countries. The apparent yearly per capita consumption was around 10–11 kg during the '70s, compared to 18 kg since 1995 (ISTAT, various years). Demand for fresh or frozen fish is also relevant at 14.4 kg per capita. The 2001 at home consumption of fish is 7.8 kg/pc/year, that is 450,000 metric tons (1.2% less than in 2000) and 3,648 million EUR (8% of total food expenditure). The (unprocessed) fresh marine fish at home consumption is 1,078 million EUR (131,166 t) (ISMEA-Nielsen, 2001b, 2002)[2]. Both the actual average level of at home fresh or frozen fish demand and the rate of households consuming it (76%) are due to: a) gradual decline of traditional regional patterns (increasing consumption from north-west to south Italy); b) increasing attention to diet and the relation between food and health; c) the substantial stabilization or decrease in fish retail nominal prices over time; d) increasing market share of large scale retail (28.3% of total fresh fish demand in 1997, 40.2% in 2001) (ISMEA, 2001c, Uniprom, 2001).

Until recently, households had little accurate information about the fresh fish consumed in regards to its country of production or catch, its origin (wild or farmed) and the production process (extensive or intensive). As a consequence of asymmetric information, prices did not provide adequate signals of quality and/or origin. An improvement is expected by the mandatory EC Regulation 2065/01, applied in January 2002, on informing consumers about the method of production and catch area of fishery and aquaculture products.

On the other hand, in recent years demand for organic food has exponentially grown. In 2001, ISMEA-Nielsen estimated that the total household demand was 1,174 million EUR, equal to a 2.7% share of total food expenditure, with a target share of at least 5% in 2005 (ISMEA, 2001b). The 12.8% increase in demand for organic food since 2000 can be considered the best performance amongst food sub-sectors, despite the relatively low rate of households consuming organic products (medium-high income, relatively young families, mainly living in north-west (36.7%) or north-east Italy (26.6%)). This success is due to an increasing demand for food security and environmental-friendly farming but also to modern retail trader strategies aiming at increasing their market share with product differentiation. The increasing share large scale retailers have of the organic food market is currently 23%[3] (Yussefi and Willer, 2002) (94% in quantity and 86% in value[4] (ISMEA, 2001d), taking into account only packaged organic food).

Despite the statutory EU regulation on organic food there is also confusion amongst consumers over the meaning of organic and many ask for more information and guarantees concerning certification processes (Federalimentare, 2002).

On the supply side, Italy is a net importer of fishery and aquaculture products. Excluding molluscs (72.8%) and trout (17.8%), the most important aquaculture products are seabass (2.9% of total production for 1999) and seabream (2.3%) (ISMEA, 2001a). In 1998, there were 79 marine species harvested intensive or extensive on small to medium-size farms, 19 of which located off-shore (ISMEA, 1998 and 2000). Their production has

---

[2] Consumer attitudes towards fresh fish are quite different with regard to at-home-consumption (health food) and away-from-home-consumption (mainly prestige food) (Boatto and Defrancesco, 1994; Dellenbarger et al., 1992; Trevisan, 1999).
[3] Direct marketing 17%, specialized shops 60%.
[4] Several large scale retailers adopt an organic private labelling strategy (generally highly priced) and the others prefer a lower pricing policy compared to specialized organic food retailers.

Table 1. Seabream and seabass production, import export and consumption, Italy (metric tons)

	1995	1996	1997	1998	1999	2000	2001
SEABREAM							
Production	5,379	5,393	5,759	7,217	7,454	7,939	
– catch	2,179	1,743	1,859	1,717	1,754	1,939	
– aquaculture	3,200	3,650	3,900	5,500	5,700	6,000	
% EU production	26.8	21.2	18.6	19.8	14.7	13.3	
% world production	17.4	13.8	12.2	12.0	10.0	8.3	
Imports		2,582	4,681	5,574	9,158	10,619	
Exports		605	334	434	984	1,179	
At home consumption			11,556	12,171	18,830	18,472	17,013
SEABASS							
Production	8,233	6281	6,630	7,739	9,081	10,295	
– catch	4,633	2481	2,030	1,889	1,881	2,195	
– aquaculture	3,600	3800	4,600	5,850	7,200	8,100	
% EU production	32.3	24.2	22.5	22.2	21.1	21.5	
% world production	28.4	21.3	19.1	18.5	18.5	16.8	
Imports				7,176	10,277	11,340	
Exports				156	337	537	
At home consumption			6,122	6,813	9,843	10,900	9,741

Source: ISMEA Web-database. Data not fully balanced as different sources have been used: FAO (total production in fresh fish equivalent), ISTAT (imports, exports of fresh and processed products), ISMEA-Nielsen (only fresh fish at home consumption). Blanks represent unavailable data.

increased over time, as demand for these species has increased (Table 1), assuring a regular supply during the year[5].

Over recent years the profitability of marine species (mainly seabream) has considerably dropped, showing operating losses in many semi-extensive farms. This is due to the strong price competition of imported products from foreign countries, namely Greece and, more recently, Turkey (ISMEA, 2000), where intensive off-shore farms have been established. On the other hand, the abovementioned lack of consumer information regarding product origin and the substantial price taker position of small-scale fish farms along the supply chain does not allow the adoption of product differentiation strategies, which assure to producers a price premium. An exception is origin labeling, introduced by some producers' associations located in traditional and knowledgable areas. As a consequence, the price paid to producers has decreased over time to an average of 4.6–6.7 EUR/kg for both species (small-medium size fish). Only fish over 800 g maintained a quite stable nominal price (over 10 EUR/kg)[6]. The retail nominal price dropped by around 2 EUR/kg (–20%) from 1997 to 2001.

In order to avoid this sector crisis, an unexplored product differentiation strategy for Italian sea-fish farmers could be certified organic marine fish, taking into account the increasing demand for organic food.

At present there is no supply of fresh organic fish in Italy, whether under independent certification schemes, national or EU legislation (with the exception of a few organic

---

[5] Usually a supply increase occurs in the last months of the year due to the extensive lagoon catch.
[6] For this reason, some farms, able to sustain the financial cost due to a longer production cycle, differentiated their production to large-size fish.

trout farms certified by an independent certification body). In other countries the situation is quickly changing. At the time of writing, fish was not included in the EU organic farming regulations (EC Reg. 2092/91 regards the certification of organic food labeling, amended by EC Reg. 1904/99 to cover organic terrestrial animal husbandry). IFOAM has recently begun work on international guidelines for organic aquaculture; however, several organic labeling bodies provide standard of production and labeling schemes for organic fish (mainly salmon) in several Northern-Central European countries, the USA and Canada (EU Commission, 2000; Uniprom, 2002).

This paper shows the results of an experimental multi-disciplinary research program (2000–01) aimed at testing the technical and economical feasibility of an organic farming certification standard with particular reference to seabream and seabass. It should be considered as a baseline hypothesis for an EC regulation proposal, to be further checked in the future (Uniprom, 2002). In particular, the paper highlights a) on the supply side, the estimated unit production cost differential of organically farmed fish compared to conventional products harvested in four commercial intensive or semi-extensive commercial farms; b) on the demand side, both the estimated potential demand for fresh organic marine fish and consumer willingness to pay a price premium, based on a survey carried out during the promotion sale of the experimental organic fish (around 40 t) in 40 large-scale supermarket retailers.

2. METHODOLOGY AND DATA

2.1 Production costs

To estimate the technical and economical feasibility of organic fish farming in Italian aquaculture, the organic fish farming process has been experimented in four intensive or semi-extensive commercial farms located in central-southern Italy. Farmers agreed to convert part of their ponds or open-sea cages to the organic farming experimental standard under researchers' supervision and monitoring production processes in order to fine tune the standard. Obviously, the organic production processes were fully separated from the conventional ones. Highly intensive fish farms were excluded, taking into account the low fish stock density imposed by the organic farming scheme in order to improve animal welfare and to reduce health risks and the impact of farming on the environment. The farms can be considered as representative of the different Italian semi-extensive farming systems (Table 2).

Taking into account that conversion to organic fish farming involves the whole farm's structure and management (Petit, 1999); the cost estimates have been based on a full operating cost approach (Banker and Huges, 1994). Both the production and at firm sales costs have been considered, taking into account organic certification and labeling costs that are based on a national independent body's charges (AIAB). Direct costs have been analytically monitored daily on an activity based costing system (Hilton, 1997): a) rearing units' preparation and juveniles input, b) feeding (GMO free organic feed), c) monitoring, control, water changes, etc., and d) harvesting and post harvest fish

Table 2. Experimental organic fish farming units

Farm	A	B	C	D
Production System	Open-sea cages	Earth ponds	Concrete ponds	PVC ponds
No. of units	1	2	2	2
Water capacity (m^3)	1,200	1,400	200	675
Species	Seabream	Seabream	Seabass	Seabream
Final organic fish density (kg m^{-3})	10.6	3.6	15.0	13.5
Final conventional fish density (kg m^{-3})	18.8	4.1	29.0	22.0

Source: Uniprom (2002).

processing. Direct activity cost has been based on real quantities and standard average prices[7] (Selleri, 1990). Indirect costs, equipment depreciation included, have been charged to rearing units using specific cost drivers (Atkinson et al., 1998) in order to take into account the influence of low density farming on fish unit cost. The cost of organic fish has been compared with that of conventional fish by parallel-monitoring costs sustained in dimensionally comparable farm units devoted to conventional fish farming. Fish unit cost estimates have been carried out in the following stages:

1. Experiment costs evaluation. Only the grow out phase up to the commercial size of the production process has been analytically monitored, excluding larvae and fry phases. The main differences from conventional fish husbandry are due to: a) direct input farming costs, b) lower farming density, c) high monitoring and control costs, partly due to experimental farming and to reduce health risks, d) organic labeling and certification, and e) lack of farm scale-economies due to the small scale experiment, imposing high labor input.

2. Normal medium-run costs estimate. In order to better compare organic fish farming costs with those of conventional farming, normal medium-run production costs have been estimated during focus groups among researchers and farmers involved in the experimental phase. In particular, the scale-economies involved by extending the organic process to the whole farm (Monden and Hamada, 1991) have been taken into account. An example is using the feeding and monitoring systems already adopted for conventional farming, if allowed by the organic guidelines, and managing, where possible, the maximum final density per cubic meter of water graduating the harvest over time and rationalizing the over-monitoring process due to this experimental phase. The cost estimate has been extended to the whole rearing process[8].

3. Opportunity-cost. In order to evaluate the feasibility of converting existing farms to organic farming it is necessary to consider the operating results loss due to the relevant reduction of total production imposed by lower fish density. Given the actual structure and production capacity of the fish farms and

---

[7] Ad hoc organic feed has been prepared by a commercial company and supplied at a contracted price (on average 15% higher than the cost of conventional feed).

[8] Extending the analysis to the whole rearing process, organic fish production costs estimates are better in farms A and B, producing small commercial-size organic fish, and in farm D, organically rearing wild caught lagoon fish. In farm C estimated costs can be referred to the conversion phase and only with cautions to the organic farming process.

assuming the actual utilized capacity as optimal both in conventional and in organic fish farming, the opportunity cost for organic fish can be estimated. That is the differential assuring equal total operating income for the farm both from organic fish and conventional fish. The estimated opportunity cost has been expressed as operating results loss from conventional farming per kg of organically farmed fish. A sensitivity analysis of unit full production cost (including opportunity cost) to small increases in final organic fish density and to conventional fish price variation has been carried out. The variable and fixed structure of medium run costs and more optimistic market scenarios compared to those observed in recent years have been taken into account.

2.2 Consumer survey and data

As has been previously highlighted, a consumer survey has been carried out by a market research company during the promotion sale of the experimental organic marine fish harvest. The promotion was held in 40 large-scale retail supermarkets, located all over Italy, except the Northwest, during the last two weeks of November 2001 (Thursday, Friday and Saturday). The survey-scenario is as follows: a) a detailed description of an organic marine fish experimental production system was presented to consumers by the interviewer; b) the organic fish was sold at the same price as the conventional fish sold by the retailer on the same day; c) it was made clear to the consumer that the organic fish sold could be considered a potential organically-labeled new product, obtained at a higher cost of production than the existing conventional fish and therefore, be eventually supplied in the future at a higher price.

A questionnaire based personal interview was conducted on a sample of people interested in the promotion sale whether they bought the fish or not (6,877 questionnaires were collected). Due to the survey design, the results only represent the increasing number of both fresh fish and organic food buyers in large-scale supermarkets and not Italian households in general. In other words, the goal of this research is to describe the behavior of the potential 'innovator-consumer' of organic fresh fish entering the new niche market, this being the more interesting target-consumer for farms potentially converting to organic fish. From a statistical point of view, a two stage sampling technique was carried out: in the first-stage a non-random sample of supermarkets was selected in order to be spatially representative, under the promotion contracts constraints; in the second stage, household samples were based on an intercept-sampling technique (Brasini *et al.*, 1996). The questionnaire was previously fine-tuned by focus groups and its core-part was extensively pretested during a survey on household conventional fresh fish consumption carried out by Uniprom in the spring-summer of 2001 (Uniprom, 2001). In particular, pretesting helped define the range of price-increases to be randomly proposed to consumers. The 15-question interview aimed at:

a. defining household consumption patterns with regard both to fresh fish (frequency and level of seabream and seabass consumption, percentage of large-scale retail demand for fresh fish) and to organic food;

b. eliciting consumer willingness to pay (WTP) a price premium for organic fish expressed as a percentage increase on the promotional price[9]. Given the survey's protocol, the premium price can be considered a household's subjective valuation of the perceived or expected difference in quality of labeled organic marine fish compared to conventional fish (Romani, 2000). Percent WTP was first asked on the basis of a single-bounded dichotomous choice contingent valuation approach[10] (SB-CVM) (Mitchell and Carson, 1989). The take-or-leave-it percentage proposed to each respondent was randomly selected in the 0–100% range of the promotion price.

c. estimating a household's potential demand for organic marine fish (willingness to buy—WTB). The respondent's answer of "yes" or "no" to the first question was followed by a continuous follow-up question to obtain this information[11].

d. better understanding zero-WTPs, in order to separate (and exclude from the analysis[12]) protest answers from zero-WTPs expressed by potentially-in-the-market households. This understanding, along with point e, was found through an open-ended question directly asked to consumers about the percent of actual fresh seabream-seabass consumption would be potentially moved to organically farmed fish if EU-labeled organic fish were to be supplied in the future at the previously declared %WTP price increase[13].

e. describing organic marine fish innovator-households from a socio-economical point of view (gender, age, number of components, income level, region of residence, etc.).

Under Hanemann's (1984, 1989) well known linkage between random utility maximization and the functional form of econometric models with a binary dependent variable, a logit model has been estimated on SB-CVM data. It explains the log-odds ratio as a linear function of several household attributes (including income level as a covariate) and of the percentage premium price proposed (Franses and Paap, 2001; Gourieroux, 2000). Median WTP and truncated mean WTPs (both only at zero and between zero and 100%) have been calculated according to Hanemann and Kanninen (1996)[14].

Because of the appreciable number of zero answers, both on continuous WTP follow up and on open-ended WTB, a censored Tobit model has been estimated in both cases (Greene, 2000). The first one explains the declared percentage premium price as a linear

---

[9] It was preferred to express a percentage price premium instead of an absolute value, taking into account the different promotion prices among supermarkets, being set equal to prices fixed for same-size conventional fish.

[10] CV approach is been largely applied both in valuing non-market goods and in estimating premium price for new or differentiated market goods.

[11] SB-CV method suffers "yes-saying" bias, generally avoided using a multiple-bounded CV approach (Hanemann and Kanninen, 1996; Bishop and Romano, 1998). In this specific case a continuous follow up was preferred a) to evaluate WTP for a potential market good, b) in order to simplify the interviewers' work, being members of a promotion sales agency without experience in CVM.

[12] Respondents showing a low attention level during the interview (subjectively valuated by interviewer) have also been excluded. It has to be pointed out that the refusal to answer rate decreases as the attention level increases. More generally, the number of observations varies among the different analyses carried out, cases with missing data in the considered variables being excluded.

[13] Given the substantially stable household demand for fresh marine fish, it was assumed that the demand for organic fish will substitute a similar quantity of conventional fish.

[14] Parameters have been estimated by LIMDEP likelihood function maximizing routine (Greene, 2000).

function of a vector of household attributes, including the income co-variate. The second explains the potential household's monthly demand, i.e. the actual marine fresh fish consumption shift to organically farmed fish, as a linear function of income, price (premium price included), and other significant household attributes (Blend et al., 1999)[15].

3. FINDINGS AND COMMENTS

3.1 Production costs

The analysis of the estimated medium-run normal unit cost of production can be summarized as follows (Table 3)[16]:

a. Comparing the full unit operating cost of conventionally farmed fish with the average price received by producers, the operating loss incurred by the more extensive farms (A and B) due to the high price competition from imported products[17] can be highlighted. Relatively better results can be found in the more intensive farm (C) and in D, where they adopt a fish-size differentiation and a producers' association geographical origin labeling strategy.

b. The normal production costs of organic fish are 20–30% higher than for conventional fish. The main differences are directly related to the reduction in farming density and more marginally to the higher feed and monitoring costs. Feeding cost differentials are due to organic feed and to not allowing fully automated feeding equipment. Higher costs both in absolute terms and as differentials compared to those for the conventional product are sustained by more capital-intensive farms. Actually, the latter have incurred higher pond/cage-related fixed costs and indirect equipment costs (Jolly and Clonts, 1993). In other words, the maximum final organic stocking density, tentatively fixed by the pilot standard at 15 kg m^{-3}, can be considered the highest drawback to converting existing fish farms to organic farming. However, this fixed limit could be managed in a more flexible way by being more strictly related to farming conditions and by using equipment that assures animal welfare.

c. The normal costs increase compared to experimental ones, due to extending the organic standard to the whole rearing process. Although it is highly compensated by the technical and organizational scale-economies incurred by extending the organic husbandry at the farm level, mainly in the more capital-intensive farms.

---

[15] Only the selected models on the basis of the maximized value of the log-likelihood function will be later discussed.
[16] A more detailed analysis of costs can be found in Uniprom, 2002.
[17] As a result, both unaccounted family labour and accounted non monetary costs, i.e. depreciation, are generally undervalued. At present the more extensive fish farm (B) does not operate.

Table 3. Medium-run normal unit costs of organic fish and at farm price premium assuring equal operating results to conventional fish farming (EUR/kg)

Farm:	A	B	C	D
Organic fish full cost (experimental)	6.37	7.37	10.10	9.08
Organic fish full cost (normal)	6.52	7.38	7.77	7.54
Direct rearing costs	3.63	4.98	5.91	5.78
Rearing units' preparation and juveniles	1.68	1.50	3.25	2.11
Feeding	1.56	1.87	1.23	1.76
Monitoring and control	0.39	1.61	1.43	1.91
Harvesting and post-harvesting fish processing direct costs	0.78	0.69	0.55	0.71
Direct depreciation costs	0.95	0.85	0.93	0.51
Indirect costs	1.08	0.82	0.35	0.48
Certification and labeling	0.07	0.04	0.03	0.07
Conventional fish full cost	5.02	6.16	5.96	6.05
Cost difference (%)	29.7	19.8	30.5	24.6
Cost difference	1.49	1.22	1.82	1.49
Of which: direct rearing	0.71	0.67	1.02	0.91
Direct harvest and post harvest	0.42	0.39	0.39	0.31
Depreciation and indirect costs	0.29	0.12	0.38	0.20
Certification and labeling	0.07	0.04	0.03	0.07
Average price of conventional fish	5.62	5.22	7.75	8.29
Organic fish price assuring total operating results equal to conventional fish farming	7.59	7.38	10.40	10.10
Of which: normal full cost	6.52	7.38	7.77	7.54
Unit opportunity cost	1.07	(*) 0.00	2.63	2.56
Minimal price premium	1.97	2.16	2.66	(**) 1.80
Price differential (%)	34.95	41.38	34.27	21.72

(*) In this farm, showing an accounting operating loss, the analysis has been carried out with reference to break-even organic price.

(**) In this case, the minimal price premium has been reduced because of the higher average weight increase (appreciable in terms of commercial size) of organically farmed fish compared to conventional fish.

    d. Taking into account the estimated opportunity-cost of organic fish farming, the estimated price premium for organic fish, guaranteeing the firm the same total operating income as for the conventional product, varies on average from 2.07 EUR per kg in extensive farms to 2.5 in the more intensive ones. Obviously these average premium prices have to be considered as minimal values assuring only the current low profitability to farms, which is not sustainable in the long run. As previously highlighted for production costs, the minimal price premium is sensitive to the final density of organic fish stock. For example, in farm A, reaching an organic final density of 15 kg would cause the break-even price to be 6.5 EUR/kg (15% less than at experimental density). A –0.42 average elasticity of break-even price to density has been estimated. For farm B, representative of traditional Italian fish farming plants, at present out of market organic fish farming could be a new alternatiave way to operate. For the usual 4 kg m^{-3} density, the break-even price would be 7 EUR. In the more intensive farms, increasing the maximal density to 18 kg m^{-3} will cause a lesser decrease (8–10%) in the break-even price because of the higher impact of fixed and semi-fixed costs (respectively –0.55 and –0.48 average elasticity to density increase). On the other hand, if EC Reg. 2065/01 on informing consumers about fishery and aquaculture products successfully works, increasing the prices of Italian fresh sea-fish, the minimal price premium for organically farmed fish will linearly increase. On average, a 60% marginal effect can be observed in the more extensive farms as opposed to 70% in the semi-intensive ones.

3.2 Respondents' profile

Table 4 shows some summary statistics of the respondents. Households interested in the experimental organic fish promotion sale are mainly: a) residents in Northeast Italy (50.6%), b) women (60.3%), c) usually fresh fish buyers for the family (80.5%), d) 35–54 years old[18]. Both income-level distribution and family size are comparable with ISMEA-Nielsen fish-consumer surveys.

The respondents' consumption patterns have also been investigated with particular regard to organic food and to fresh marine fish. They show the expected spatial differences (Table 5). Thirty-four percent of the interviewees do not buy organic food at all, but 16% of them are regular organic food consumers, mainly living in Northeast Italy. As expected, both the market penetration and consumption level of organic food decline from the north to the south of Italy. However, this survey confirms the above mentioned need to better inform consumers on organic food production processes and on certification systems in order to increase its penetration level[19] (EU Commission, 2000; ISMEA, 2000).

The sample of respondents shows: a) an above national average marine fresh fish consumption (5.6 kg per capita year), increasing from the north to south of Italy because of traditional eating habits, b) a higher household rate of consumption, c) a relevant consumption frequency (at least weekly in over 70% of cases, as opposed to the 53% reported by ISTAT (2000), d) an above average demand for fresh fish to large-scale-retailers (67.6%). Thrity-nine point three percent of respondents buy fresh marine fish exclusively at large-scale supermarkets (51% Northeast Italy). The declining share from North to South is mainly related to differences in the rate of diffusion of the modern retail trade.

The respondents' profile confirms the hypothesis that the potential labeled organic fresh fish buyers could be, at first, both organic food and intensive marine fresh fish consumers. Caution in extending the survey results to all Italian consumers has also been confirmed.

---

[18] ISMEA-Nielsen household surveys account for a higher fresh fish consumption in higher age classes. Our findings show greater interest for organic fish consumption in relatively younger people, and confirm the general low penetration of fresh fish consumption among young families.

[19] Only 22% of respondents do not demand organic food because of its higher price.

Table 4. Respondents' summary statistics

Item	Class	%	Item	Class	%
Place of living	North-east	50.6	Age group	18–24	3.4
	Centre	18.5		25–34	13.8
	South	14.6		35–44	25.2
	Islands	16.3		45–54	25.3
Income level	low	47.8		55–64	19.9
	low-medium	26.9		>64	12.4
	medium-high	10.0	Household size	Mean value	2.9
	high	15.3	Respondents' attention level	low	9.5
Gender	female	60.3		sufficient	35.4
	male	39.7		good	55.1

Table 5. Respondents' consumption patterns: organic food and fresh fish

	Northeast	Central	South	Islands	Italy
Organic food demand (%):					
never	27.9	35.9	48.4	40.2	34.3
occasionally	51.3	49.3	45.1	49.8	49.8
regularly	20.9	14.8	6.5	10.0	15.9
Why zero organic food demand (%)					
lack of product awareness	23.9	22.3	24.2	44.1	27.3
lack of trust over certification systems	23.3	24.1	16.6	15.7	20.7
not interested	28.2	31.3	38.7	22.9	30.0
higher prices	24.6	22.3	20.6	17.3	22.0
Marine fresh fish average consumption (kg/per household /monthly)	1.06	1.31	1.57	2.08	1.39
Percentage of households not consuming fresh marine fish	25.8	22.5	9.9	9.7	20.3
Percentage of low fresh fish consumers (monthly or less)	13.4	6.4	5.7	7.8	10.0
Average large-scale-retailers share of marine fresh fish demand (%)	77.4	68.4	51.2	55.9	67.6
Average promotional price (€/kg)	9.05	6.33	6.52	6.55	7.60

3.3 Respondents' willingness to pay

Fourty-three percent of interviewed households accept the SB-CVM randomly proposed premium price. The percentage accepting to pay declines according to the increase in proposed amounts (ranging from 71.6% for a premium less than 10% to 16.6% for a price increase of over 80%). The maximum likelihood SB-logit model estimates are reported in Table 6 and have the expected sign. In particular, respondents behave according to economic theory: as the percentage price premium increases, their likelihood to accept the proposed amount decreases and the latter is positively related to income level. The coefficients on the considered consumption patterns are all positive, as expected: frequency of marine fish consumption, consumption level of both fresh fish and organic food, demand share to large-scale-retail, and traditional fish consumption habits (spatially differentiated and increasing from northern to southern Italy). Negative coefficients of household size and respondent's age are generally found in the case of eco-labeled food (Wessels et al., 1999; Blend and Van Ravenswaay, 1999; Fu et al., 1999; Jaffry et al., 2000; Asche et al., 1999) and they are consistent with Italian organic food consumption patterns. The median percentage premium price is 37.8% higher than the promotional price. The zero-truncated mean is 43.3% and the double-truncated mean 41%.

Continuous follow up WTP shows 17.7% of respondent declaring a WTP = 0. A debriefing question (Table 7) highlights that only 12% of zero-WTPs can be accounted for

Table 6. Estimated coefficients of SB-CVM logit model

Variable	Coeff.	t-value	Variable description
Constant	−1.468	−6.558	
INCOME	0.084	2.704	Income level, categorical (4 levels)
SUPER	0.005	4.494	Large scale retailers fresh marine fish demand share
FAM	−0.047*	−1.715	Family size
AGE	−0.051*	−1.898	Household age
QUA	0.095	2.935	Frequency of fresh fish consumption, categorical (5 levels)
MARINE	0.073	2.238	Monthly household fresh marine fish consumption
BID	−0.039	−26.023	Percentage premium price proposed
BIO	0.973	18.507	Household consumption of organic food, categorical (3 levels)
REGION	0.233	7.290	Place of living, categorical (4 levels)

* Significant at the 10% level; otherwise significant at least at the 5% level.
N = 5141   Percentage of correct predictions = 71.2%   McFadden's R^2 = 0.18   Model chi-square = 1257.4 (9df)

Table 7. Debriefing question on zero-WTP

	Percentage
Protest answer	11.9
In-the-market-zero (premium = 0)	46.4
Out-of-market (no concern in organic fish)	25.7
Out-of-market (lack of trust over certification)	16.0
Total	100.0

Table 8. Continuous follow-up Tobit model estimated coefficients

Variable	Coefficient	t-value
Constant	−17.67	−5.36
INCOME	0.86*	1.72
SUPER	0.12	6.80
AGE	−0.72*	−1.65
QUA	1.13	2.16
MARINE	0.94*	1.77
BIO	13.22	15.77
REGION	3.04	6.01
σ	27.66	67.64

* Significant at the 10% level; otherwise significant at least at the 5% level.
N = 2937, 17.6% of which corresponds to zero-WTP   Max log-likelihood value −8465.48

protest answers. On the other hand, 46.4% of the unwilling to pay respondents can be considered in the organic fish market, i.e. potential organic fish buyers willing to pay the same price paid for conventional fish.

Taking into account only the in-the-market households, the censored Tobit model estimates (Table 8) confirm the previous model coefficient signs of main factors affecting a WTP price premium. In particular, the role played by organic food and fresh fish direct knowledge is confirmed[20], as a consequence of an expected improvement in the perceived fish quality. As expected, the censored conditional mean household percentage premium price (29.46%) is lower than the SB-CVM mean, because of the "yes-saying" bias affecting the latter.

From the expressed household premium price for organically-labeled fish in EUR/kg (Figure 1), it can be highlighted that:

a. innovator-potential-consumers of organically farmed fresh marine fish are willing to pay 2.25 EUR/kg as a mean premium price ($\sigma = 1.98$). If fully transferred along the *filière* to organic fish farmers, the mean premium is over the minimal premium at the farm level only in the case of more extensive fish-farms. On the other hand, the result seems too low, in general and particularly for the more intensive fish farms, taking into account respondents' tendencies to overestimate their real WTP in simulated markets (Romani, 2000; Dalli and Romani, 2000). For the last fish farm considered, mainly producing origin-labeled large-sized fish, the effect of a double-labeling system seems questionable in terms of price premium, and further investigations are needed[21].

b. Mean premium price significantly differs among respondent subgroups. In particular, it is higher and more adequate for organic fish farms' needs; in the case of regular organic food consumers, it expresses a high level of marine fish demand for consumers who live mainly in Northern Italy. It seems a relatively easy and "expert" market segment for organic fish farm penetration strategies, because relatively low investment costs are needed in order to inform consumers of organic fish farming standards.

---

[20] In another model specification, including the promotion price, the coefficient was not significantly different from zero, being income-related.

[21] In all cases, the experimental organic fish was unlabeled during the promotion sale.

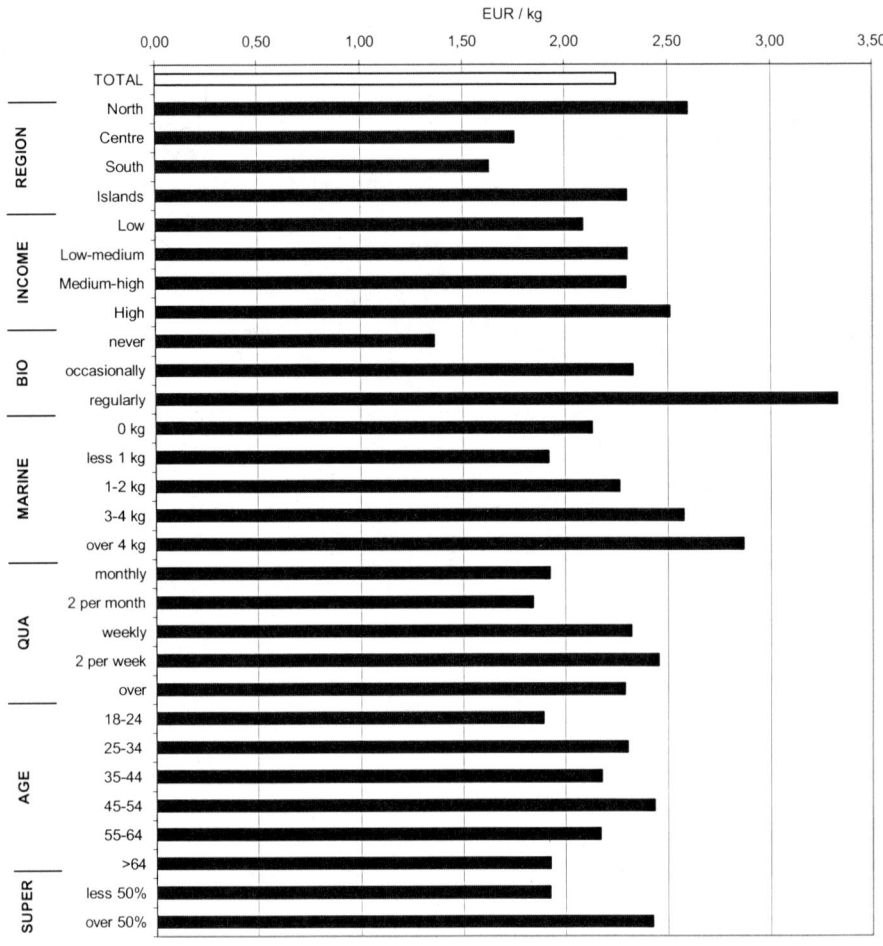

Figure 1. Average household premium price (EUR/kg)

3.4 Respondents' potential demand

Taking into account only the potentially in-the-market households, a tentative[22] censored Tobit linear demand model has been estimated (Table 9) expressing the dependent variable as potential at-home organic marine fish demand as a household's monthly consumption (kg). It has been obtained on the basis of declared %WTB and actual conventional marine fish at-home consumption. As expected, both the decision to buy and the

---

[22] A further data cleaning was needed at this stage of the analysis, in order to exclude both WTB missing data and the few cases expressing an inconsistent behaviour (WTP > 0 but WTB = 0). There are 1710 remaining valid cases.

Table 9. Tobit demand model estimated coefficients

Variable	Coefficient	t-value
Constant	−0.698	−6.174
INCOME	0.045	2.591
ORGANIC PRICE	−0.012*	−1.657
MARINE	0.459	23.359
BIO	0.060	1.980
REGION	−0.125	−6.899
σ	0.759	49.523

* Significant at the 10% level; otherwise significant at least at the 5% level.
N = 1710, 21.2% of which correspond to zero-WTB   Max log-likelihood value −1805.89

quantity purchased (Tobit model imposing that the same variables affect the two decisions) are positively related to a household's income level and negatively to price. Mean price elasticity of demand equals −0.22, showing a scarce price sensitivity of innovator-potential-buyers. On the other hand, both organic food and marine fish consumption levels show a significant positive effect on demand. However, the organic fish demand is more relevant in Northern Italy, probably as a consequence of: a) higher trust in the organic labeling system, and b) expected reduction in imperfect and asymmetric information of marine fish consumers due to organic certification. The need to be better informed on marine fish origin and quality is actually higher in northern regions, where at-home marine fish consumption has recently become more widely spread.

The household's estimated monthly mean potential demand is 511g, around 37% of the respondents' actual marine fish consumption. It is an interesting level (2.1 kg on a per capita year base), from the producers' point of view and with a little skepticism, because of the particular innovator-consumers considered. The mean potential demand for organic marine fish consumers is above average in several homogeneous subgroups: high organic food consumers (728 g), high income level (673 g), 45–55 year old households (605 g), high marine fresh fish consumers (1.281 kg) and families living in Northern Italy (543g). The latter expresses a relevant %WTB (43.6%) which is comparable with that of frequent organic food consumers (44.4%).

4. CONCLUDING REMARKS

Analysis has shown the potential interest in organic farming labeling as a means of generating market driven incentives to support Italian marine-fish aquaculture. On the demand side, the consumer survey carried out during the promotion sale of experimental organically farmed marine fish has shown:

a. the potential innovator-consumer is willing to pay an average premium price, able to cover the increased production costs for organic standards, at least, of more extensive marine fish farms. On the other hand, a well defined household subgroup (high-income level, regular organic food consumer, above average fresh marine fish buyer, mainly living in Northern Italy) provides for an intentional WTP a premium fully compatible with the estimated costs at the farm level;

b. the potential demand, cautiously estimated as a percentage shift from conventional consumption levels to organically labeled consumption is interesting in volume if compared to Italian aquaculture marine fish supply.

At the marine fish farm managerial level, the following are of note:

a. a product differentiation strategy based on organic fish labeling seems possible in the short run. It aims to reduce the loss of profitability recently incurred by Italian marine fish farmers because of the strong price-competition of imports. The previously described target sub-population of consumers seems a relatively-easy market segment for organic fish farm penetration strategies due to the relatively low investment costs needed to inform consumers on organic fish farming standards. At present, this target is the best informed on organic food certification systems. In the meantime, organic labeling seems to play the role of improving perceived fish quality and reducing the consumers' lack of information on fish origin (Roth et al., 2000);
b. since fish farmers operate as price takes, the full transferability of the consumers' price premium to them seems questionable. Adequate supply concentrations at the farm level as well as price agreements with large-scale retailers are strongly recommended.

At the institutional level, the following are of note:

a. if the niche market seems to adequately assure profitability to organic marine fish farms in the medium run. Temporary institutional financial support is needed to cover the higher investment and organizational costs sustained by fish farms during the conversion period when unlabeled fish is produced. A temporary and partially decoupled support system could be arranged on the rearing capacity, such as the EU agriculture organic farming support system under the Agenda 2000 Regulation. The lower environmental impact of organic aquaculture could justify the support (Cahill, 2001);
b. in order to assure an increase in demand for organically labeled marine fish and organic food in general, consumers have to be better informed on organic farming guidelines to improve their trust in certification systems (Marette et al., 1999);
c. the imperfect and asymmetric information characterizing the fresh fish market could also be reduced. Under EC Reg. 2065/01 on informing consumers about fishery and aquaculture products, an EU member country can autonomously define both the different production/catch processes and the farming/catch areas. In Italy, a proper fine-tuning strategy could increase both fresh fish demand and consumer willingness to pay for origin-labeled or organic-labeled fish.

Finally, further research needs have emerged. From the supply side, the pilot organic standard has to be improved, both from a technical and economical point of view, to give greater attention to the whole production process costs and the distribution costs. The *interfilière* price transfer mechanism also has to be explored. On the demand side the following should be investigated: a) the double labeling system effect (geographical origin label and organic label) on the price premium, under a full operating EC Reg. 2065/2001; b) the price premium effect for different species and different fish commercial sizes; c) potential away-from-home demand for certified organic marine fish because of the expected higher consumer WTP affecting it.

5. REFERENCES

Asche, F., Bremnes, H., and Wessel, C.R., 1999, Product Aggregation, Market Integration and World Salmon Prices, *American Journal of Agricultural Economics*, **81**(3): 568–581.
Atkinson, A.A., Banker, R.D., Kaplan, R.S., and Young, S.M., 1998, *Management Accounting*, ISEDI, Torino.
Bishop, R.C., and Romano, D. (eds.), 1998, *Environmental Resource Valuation: Applications of the Contingent Valuation in Italy*, Kluwer Academic Publishers, Boston.
Blend, J.R., and Van Ravenswaay, E.O., 1999, Measuring consumer demand for ecolabeled apples, *American Journal of Agricultural Economics*, **81**(5): 1072–1077.
Boatto, V., and Defrancesco, E., 1994, *L'economia ittica in Veneto: dalla produzione al consumo*, ASAP, Venezia.
Brasini, S., Tassinari, F., and Tassinari, G., 1996, *Marketing e pubblicità: metodi di analisi statistica*, Il Mulino, Bologna.
Banker, R.D., and Huges, J.S., 1994, Product Costing and Pricing, *The Accounting Review*, **7**: 479–494.
Cahill, C., 2001, The multifunctionality of agriculture: what it means?, *Euro Choices*, **1**: 36–40.
Dalli, D., and Romani, S., 2000, *Il comportamento del consumatore: teoria ed applicazioni di marketing*, Franco Angeli, Milano.
Dellenbarger, L.E., Dillard, J., Shupp, A.R., Zapata, H.O., and Young, B.T., 1992, Socioeconomic factors associated with at-home and away-from home catfish consumption in the United States, *Agribusiness*, **8**(1): 35–46.
EU Commission, 2000, *Organic salmon production and Consumption: Ethics, Consumer perceptions and regulation (ORGSAL)*, FAIR-CT98-3372, Report to EU Commission DG XII.
Federalimentare, 2002, *L'industria alimentare italiana nel 2001, Rapporto alla assemblea annuale*, Parma.
Franses, P.H., and Paap, R., 2001, *Quantitative models in marketing research*, Cambridge University Press, Cambridge.
Fu, T., Liu, J., and Hammitt, J.K., 1999, Consumer willingness to pay for low-pesticide fresh produce in Taiwan, *Journal of Agricultural Economics*, **50**(2): 220–233.
Gourieroux, C., 2000, *Econometrics of Qualitative Dependent variables*, Cambridge University Press, Cambridge.
Greene, W.H., 2000, *Econometric Analysis*, 4th Edition, Prentice Hall International, Upper Saddle River.
Hanemann, W.M., 1984, Welfare evaluations in contingent valuations experiments with discrete responses, *American Journal of Agricultural Economics*, **66**(3): 332–341.
Hanemann, W.M., 1989, Welfare evaluations in contingent valuations experiments with discrete response data: Reply, *American Journal of Agricultural Economics*, **71**(4): 1057–1061.
Hanemann, W.M., and Kanninen, B., 1996, *The statistical analysis of discrete-response CV data*, Department of Agricultural and Resource Economics, University of California at Berkeley, Working paper n. 798.
Hilton, R.W., 1997, *Managerial Accounting*, McGraw-Hill, New York.
ISMEA-Nielsen, 2001a, *Acquisti domestici di prodotti ittici in Italia (1997–2000)*, Roma.
ISMEA–Nielsen, 2001b, *Gli acquisti domestici di prodotti ittici: 8/4-5/5/01*, Roma
ISMEA-Nielsen, 2002, *Gli acquisti di prodotti ittici delle famiglie italiane. Rapporto nazionale e per circoscrizioni geografiche* (draft), Roma.
ISMEA, 1998, 2000, 2001a, *Filiera pesca e acquacoltura*, Roma.
ISMEA, 2001b, *La spesa alimentare di prodotti biologici*, Roma.
ISMEA, 2001c, *Le tendenze del settore agroalimentare nel 2000*, Roma.
ISMEA, 2001d, *Indagine sui prodotti biologici presso la GDA*, Roma.
ISTAT, various years, *Indagine sui consumi delle famiglie*, Roma.
ISTAT, 2001, *Annuario Statistico italiano*, Roma.
Jaffry, S., Pickering, H., Wattage, P., Whitmarsh, D., Frere, J., Roth, E. and Nielsen, M., 2000, *Consumer Choice for Quality and Sustainability in Seafood Products: Empirical Findings from United Kingdom*, IIFET 2000 Conference proceedings, Corvallis, Oregon
Jolly, C.M., and Clonts, H.A., 1993, *Economics of aquaculture*, The Haworth Press, New York.
Marette, S., Crespi, J.M., and Schiavina, A., 1999, The role of common labelling in a context of asymmetric information, *European Review of Agricultural Economics*, **36**(2): 167–178.

Mitchell, R.C., and Carson, R.I., 1989, *Using Surveys to Value Public Goods: the Contingent Valuation Method*, Resource for the Future, Washington D.C.

Monden, Y., and Hamada, K., 1991, Target Costing and Kaizen Costing in Japanese Automobile Companies, *Journal of Management Accounting Research*, **3**: 16–34.

Petit, J. (ed.), 1999, *Environnement et aquaculture*, INRA Ed., Paris.

Romani, S., 2000, *L'analisi del comportamento del consumatore per la determinazione del prezzo di vendita di prodotti e servizi*, Franco Angeli, Milano.

Roth, E., Nielsen, M., Pickering, H., Jaffry, S., Whitmarsh, D., Wattage, P., and Frere, J., 2001, The price of fish quality, *IIFET 2000 Conference proceedings*, Corvallis, Oregon.

Selleri, L., 1990, *Contabilità dei costi e contabilità analitica*, EtasLibri, Milano.

Trevisan, G. (ed.), 1999, *Il prodotto ittico: consumo, qualità, commercializzazione*, Università Ca'Foscari, MIPAF, Venezia.

Uniprom, 2001, *Indagine quanti-qualitativa sui consumi di prodotti ittici rivolta alla clientela dei punti vendita appartenenti ad organizzazioni della distribuzione organizzata*, mimeo, Roma.

Uniprom, 2002, *Verso l'acquacoltura biologica?*, Roma.

Wessels, C.R., Donath, H., and Johnson, R.J., 1999, Assessing consumer preferences for ecolabeled seafood: the influence of species, certifier and household attributes, *American Journal of Agricultural economics*, **81**(5): 1084–1089.

Yussefi, M., and Willer, H., 2002, *Organic Agriculture Worldwide 2002*, Biofach, 74, Nürnberg.

ITALIAN CONSUMERS' PREFERENCES AND WILLINGNESS TO PAY FOR ORGANIC BEEF
A Survey in Piedmont

Alessandro Corsi and Silvia Novelli[*]

SUMMARY

In this paper the results of a study aimed to assess prospective consumers' behavior concerning organic beef, with particular emphasis on consumers' willingness to pay for it, are presented. The study is based on a random telephone survey conducted in 2001, in the Piedmont Region. The results suggest that a large share of consumers are willing to buy organic beef even at prices appreciably higher than prices of regular meat, and the maximum prices at which they are prepared to purchase organic meat are quite high. Preferred selling modalities mainly depend on present purchasing habits. Organic beef directly cut by butchers would be preferred to packaged and labeled beef. Organic beef might therefore gain an appreciable market share. The reasons for its present scarce consumption are probably due to the supply side of the market.

1. INTRODUCTION

This paper reports the results of a research project funded by the Piedmont Region and promoted by Agri.Bio Piemonte, a regional association of organic farmers. The project aimed to assess consumers' attitudes and willingness to pay (WTP) for organic beef. The latter was the main point of interest, but also consumers' preferences about selling modalities and points of sale and familiarity with organic produce were investigated. The project was developed during 2001 and finished in 2002.

For a long time, certification of organic produce in Italy was left to private labels, typically belonging to organic producers' associations. Council Regulation (EEC) 2092/91 dictates the characteristics that organic products should comply with to be labeled "organic". Nevertheless, this regulation concerned vegetal products and left out

---

[*] Department of Economics "S. Cognetti de Martiis"—University of Turin, and Centro Studi per lo Sviluppo Rurale della Collina (Research Center for Rural Development of Hilly Areas), respectively.

animal products. In 1999, a Regulation concerning these products (Council Regulation (EC) 1804/1999) was issued. The national regulation concerning animal products was passed in 2000 (D.M. 4/8/2000, n. 91436). In short, Italy could not legally market organic animal products with legally guaranteed labels before the year 2000. Given the legal protection ensured by the regulation, organic producers were interested in knowing the consumption prospects of organic beef. The first research on this issue was carried out in 2000 (Marengo et al., 2000) among the operators of the marketing chain (slaughterhouses, supermarkets, butchers, specialized organic shops) concerning their attitudes and possible problems in marketing and processing organic beef. The current study was directly investigating consumers' attitudes.

2. DATA

Since the survey was funded by the Piedmont Region and sponsored by a regional association, the reference population was the regional one. It was decided, because of budget constraints, to use a telephone questionnaire. The survey had three specific goals: a) to analyze changes in consumers' behavior after Bovine Spongiform Encephalopathy (BSE) events and consumers' knowledge and purchase habits of organic products; b) to evaluate consumers' willingness to pay for organic beef; c) to determine consumers' preferences about organic beef selling outlets, packaging and labels. These corresponded to three parts of the questionnaire, followed by questions concerning socio-economic characteristics.

The first part, after a filter question about beef consumption or the reasons for not consuming it, if this was the case, asked about changes induced by the BSE crisis and purchasing habits. Also, consumption habits of and familiarity with organic produce were investigated.

Using a closed-ended format, the central part of the interview concerned willingness to pay for organic beef. First an explanation was given about the prospective availability, the characteristics and the certification process of organic beef. Then respondents were asked whether they would pay a specific price (bid price) to buy organic beef. A follow-up question followed: those respondents who had answered 'yes' to the first question were asked again if they were willing to pay a second higher price; if the answer to the first question was 'no' the interviewer proposed a lower price. This "double-bound" format increases the elicitation process efficiency (Carson et al., 1986; Hanemann et al., 1991).

Prices differ among meat cuts and it was obviously impossible to include all cuts in the questionnaire. Therefore, it was decided to evaluate two meat cuts largely popular among Italian consumers, roast and minute steak. They are characterized by different prices and cooking processes, the former is cheaper and more time-consuming for cooking than the latter.

The elicitation question for those persons presently consuming regular meat was as follows: "Assume you can find on the market certified organic beef meat; if roast cost X ITL/kg, would you buy it?" Three answers were prompted: "Yes, I would buy it in the same quantity I'm currently consuming"; "Yes, but I would buy less than what I'm currently consuming"; "No". These respondents were also asked about the price they currently paid for regular meat.

Respondents who had answered, to a previous question, that they had given up eating beef after the "mad cow" events were asked about the possibility of going back and consuming it once again. The wording of the elicitation question in this case was: "Assume you can find on the market certified organic beef meat; if roast cost X ITL/kg, would you buy it again?" In this case, the answer could only be "yes" or "no". For these respondents the question about prices currently paid was obviously omitted.

The same questions were asked for minute steak. To avoid a question order bias, six different versions of the questionnaire were randomly submitted to the respondents, each different in the ordering of the questions and/or of the provided answers.

The bid vector of the X prices was set based on a preliminary inspection of regular beef prices. Organic beef was assumed to be more expensive than regular meat, due to higher production costs and specialized distribution. Bid prices were therefore set higher than, or equal to, first-rate quality meat currently sold. Bids were randomly submitted to the respondents. When the respondent stated willingness to pay the first bid price, the second bid price asked was 5,000 ITL/kg (2.58 EUR/kg) higher. If the respondent was unwilling to pay the first price, then he/she was asked a second one, reduced by the same amount.

The third part of the questionnaire concerned the most trusted form of selling and the preferred labeling. Finally, socio-economic characteristics were asked: number of household members, gender, education, and profession of the respondent, household income bracket.

The questionnaire was pre-tested with a small pilot sample in order to assess the adequacy of the bid design and the clearness of the questionnaire.

Data were collected through a random telephone survey in June–July 2001. The target population was those residents in the Piedmont Region who were usually in charge of purchasing food for themselves and their family. A sample of families living in the Piedmont region was randomly drawn from the electronic telephone directory[1]. A total of 879 families living in the region were contacted. Interviewers explicitly asked to speak to the household member who was responsible for food shopping. The response rate was 51.4 percent, a reasonably good share for a telephone survey. The interviewers stopped some interviews (4.9 percent) when respondents were found to be permanently out of the beef market (vegetarians, people consuming only other meat for health reasons and farmers self-consuming their products). Finally, 0.8 percent of the questionnaires were not usable because of incomplete information (respondents were unable to state their WTP). In conclusion, a final sample of 402 questionnaires was successfully completed. Part of the respondents who completed the questionnaire did not consume specifically roast or minute steak; so, the usable number of questionnaires employed to estimate WTP for organic beef was 376 for roast and 397 for minute steak.

Table 1 reports the descriptive statistics of the explanatory variables. They include respondents' socio-demographic characteristics[2] (gender, age, education, household size, household income classes), their residence (divided in small—less than 50,000 inhabitants—and big towns) and a dummy variable indicating their answer to the question

---

[1] Bias due to unlisted telephone numbers has been assumed to be marginal, since the share of households not having a telephone is very low.
[2] Since 15.2 percent of the respondents refused to reveal their family income, missing income values were inputed, regressing socio-economic variables on income for the complete questionnaires, using the estimated parameters to predict missing values, and attributing the observations to the relevant income classes.

Table 1. Descriptive statistics of the explanatory variables (402 observations)

Variable	Mean	Standard deviation
Price of regular roast (thousand ITL/kg) [*]	25.892	4.790
Price of regular minute steak (thousand ITL/kg) [*]	29.547	5.591
Big town (=1 if living in towns with more than 50,000 inhabitants)	0.311	0.463
Sex (female = 1)	0.818	0.386
Age (*years*)	50.108	15.612
Education (years of study)	10.313	3.852
Household size (number of family members)	3.189	1.052
Family income classes [**]		
0–15 million ITL/year (0–7,747 EUR)	0.080	0.271
15–30 million ITL/year (7,747–15,494 EUR)	0.308	0.462
30–45 million ITL/year (15,494–23,241 EUR)	0.338	0.474
45–60 million ITL/year (23,241–30,987 EUR)	0.194	0.396
Over 60 million ITL/year (over 30,987 EUR)	0.080	0.271
Familiar with organic products (*1 = yes*)	0.639	0.481

[*] Calculated for consumers of the specific meat cut who could remember the price.
[**] Values replaced by fitted values (see footnote 2).

whether they were familiar with organic products, which supposedly could influence their preference for organic meat. A comparison of the sample with the entire population is uncertain because the reference population only includes the people in charge of purchasing food, not a random drawing from the entire population. Nevertheless, the sample characteristics, whenever possible, were compared to Census data. In our sample, the share of women is obviously much higher, as expected, because they more frequently take care of buying food (82 vs. 52 percent); the younger age group (20–39) is slightly underrepresented (31 vs. 36 percent); the same applies to people with lower education (no respondent without any school diploma is included in the sample, while they are 6.4 percent in the Region; the relevant shares for elementary school are 19 vs. 38 percent). Inference of the results relating to the general population should therefore be done with some caution, because of possible bias.

3. WILLINGNESS TO PAY FOR ORGANIC BEEF

The results as to consumers' willingness to pay can be shown in different ways:

1. as the share of consumers willing to buy organic beef at different price levels;
2. as the share of consumers willing to buy organic beef at different price premiums, relative to the price paid for regular beef;

3. in terms of average maximum price at which consumers would buy: i) the minimum quantity of organic beef or ii) the same quantity they usually purchased of regular quality.

3.1. Shares of consumers at different prices

The shares of consumers stating they would buy organic beef at different prices are shown in Tables 2 and 3. The shares address the answers to the first bid price of those respondents who were consumers of either regular roast or minute steak.

At a price of organic *roast* of 25,000 ITL/kg (12.91 EUR/kg), 74.6 percent stated they would buy the same quantity of organic roast as they usually buy of conventional beef (Table 2). A further 19.5 percent would buy it, but in a smaller quantity, bringing the total prospective consumers of organic roast to 94.1 percent. This is a very high share, but it should be borne in mind that the average price paid for conventional roast was only slightly higher, 25,892 ITL/kg. With a price of 30,000 ITL/kg (15.49 EUR/kg), the share of those stating they would purchase the same quantity is lower, but the share increased for those stating they would purchase it in a smaller quantity. Therefore the total prospective consumers are still 91.1 percent. Only at the third prospected price, 35,000 ITL/kg (18.08 EUR/kg), does the stated consumption sensibly decrease to 79.1 percent.

The results for consumers of conventional *minute steak* are similar (Table 3). For a prospected price of 30,000 ITL/kg (15.49 EUR/kg) the share of those stating they would purchase the same quantity was 72.1 percent. Those stating they would buy it in a smaller quantity was 23 percent, for a total of 95.1 percent. For a price of 35,000 ITL/kg (18.08 EUR/kg) the shares were 62.8, 26.5 and 89.4 percent, respectively. The relevant shares for a price of 40,000 ITL/kg (20.66 EUR/kg) were 20.8 and 45.6 percent, with a total of 66.4 percent. The average price for this cut was 29,546 ITL/kg.

Table 2. Shares of consumers of regular roast stating they would buy organic roast at different bid prices (ITL/kg). N = 340

Answers	Bid price levels		
	25,000	30,000	35,000
Yes, same quantity	74.6	63.4	42.7
Yes, but less	19.5	27.7	36.4
No	5.9	8.9	20.9
Total	100.0	100.0	100.0

Table 3. Shares of consumers of regular minute steak stating they would buy organic minute steak at different bid prices (ITL/kg). N = 360

Answers	Bid price levels		
	30,000	35,000	40,000
Yes, same quantity	72.1	62.8	20.8
Yes, but less	23.0	26.5	45.6
No	4.9	10.6	33.6
Total	100.0	100.0	100.0

Non-consumers of conventional beef can obviously only be asked whether they would buy some organic beef or not at a given price. In general, the share of "yes" responses of this group is lower than the shares of consumers willing to buy some organic beef (i.e. those stating they would buy the same quantity plus those who would buy less). Nevertheless, non-consumers of regular beef are few (just 39 interviewees) and the relevant shares are therefore subject to a larger variability. In any case, if we consider the overall shares of respondents willing to buy some organic beef at different prices the general picture does not change substantially when including non-consumers (Tables 4 and 5).

The data, therefore, suggest good market prospects for organic beef. Some discussion is nevertheless required in interpreting these results, since they crucially depend on the consistency between stated preferences and future actual behaviour when organic beef becomes available. While it is impossible to totally dismiss an hypothetical bias, it is rather unlikely that it is a big problem when asking respondents to state their prospective behaviour for a consumption good, such as beef. Hypothetical bias and related biases may be a problem when evaluating environmental public goods, e.g. wild species protection schemes, with which respondents have little familiarity and for which the effort of formulating a response is high. By contrast, the interviewees are quite familiar with beef. Moreover, a consumption good such as beef has little symbolic value. The risk of a "warm glow" effect (Andreoni, 1990; Kahneman and Knetsch, 1991), a response stating the importance attached to the issue and not the real WTP, is therefore limited.

On the whole, it seems likely that a very large part of consumers would be prepared to shift to organic beef, at the prospected prices. Nevertheless, some further element can be considered. The prospected prices can be evaluated as high or low contingent on the price of regular beef for those respondents who are beef consumers. The average price of regular roast was slightly higher than the lowest bid price and the average price for regular minute steak was slightly below the lowest bid price. So, it is not surprising that high shares of consumers would prefer organic beef for prices similar to the ones of regular beef, considering that organic is considered higher quality than regular beef. Second, prices of regular beef turned out to have a much larger variation than suspected. The average price for regular roast was 25.9 thousand ITL/kg, with a standard deviation of 4.8, and a range from 12 to 35.5 thousand ITL/kg. For minute steak, the relevant data was 29.5 and 5.6 thousand ITL/kg, and the range went from 14 to 40 thousand ITL/kg. Hence, the results can be better evaluated when compared to prices paid for regular beef.

3.2. Shares of consumers at different premiums

A comparison between the price paid for regular beef and the bid price is only possible for those respondents who consumed regular beef. Nevertheless, a large number of them could not remember the price they paid; more exactly, 139 out of the 340 respondents who consumed regular roast and 132 out of the 360 respondents who consumed regular minute steak could not remember the relevant prices. This exercise is based on the remaining respondents[3]. We consider both those who responded they would have

---

[3] This may lead to a bias if the responses to the WTP question were correlated to remembering the price; the results should therefore be interpreted with some caution.

Table 4. Shares of consumers of regular roast stating they would buy organic roast at different bid prices (ITL/kg). N = 379

Answers	Bid price levels		
	25,000	30,000	35,000
Consumers	94.1	91.1	79.1
Non-consumers	86.7	76.9	27.3
Overall	93.2	89.6	74.4

Table 5. Shares of consumers of regular minute steak stating they would buy organic minute steak at different bid prices (ITL/kg). N = 399

Answers	30,000	35,000	40,000
Consumers	95.1	89.4	66.4
Non-consumers	66.7	80.0	11.1
Overall	92.0	88.3	62.7

purchased the same quantity as regular beef ("yes" responses) and those who responded they would buy organic beef but at a less amount than regular beef ("yes, but less" responses). The former is obviously a more conservative assessment of prospective consumption of organic beef.

Tables 6 and 7 present the results for roast and minute steak respectively. The difference between the bid and the actual price is divided into intervals[4]. The results show that for a price up to 30 percent over the price of regular *roast* (Table 6), the share of consumers willing to buy the same quantity as regular beef is still large. The share of "yes" responses sharply decreases for higher prices and falls to zero for price increases over 40 percent. Conversely, the share of "yes, but less" responses increases up to prices 60 percent higher than regular roast, then falls rapidly.

As for *minute steak* (Table 7), the trends are somewhat more accentuated. For this cut, the share of "yes" responses sharply decrease for prices 20 percent higher than regular beef, and the shares of "yes, but less" responses for higher premiums are lower than it is the case for roast.

These results are consistent with economic theory and also suggest the patterns of substitutability in consumption of organic and regular beef. The "yes" responses ideally refer to consumers that have a fixed consumption, and are prepared to totally substitute organic for regular beef. By contrast, those who respond "yes, but less" either are trading a higher quality for a lower quantity (if they are going to purchase only organic beef) or are combining the consumption of regular and organic beef. Of course, their decisions depend on the organic beef price. The higher the price the less total substitution

---

[4] Since the bid prices were randomly submitted to respondents, in some cases the bid price was equal or lower than the price paid for regular beef. In few of these cases (4 over 66 for roast, and 4 over 61 for minute steak) we obtained negative responses. In principle, therefore, they would pay less for organic beef than for regular beef and, hence, they evaluate the former as inferior to the latter. More realistically, this results from random errors (for instance, from respondents simply not thinking about the price of regular beef when answering the question about their willingness to pay). In fact, the shares of "no" responses for those asked their willingness to pay for organic beef prices lower than prices of regular beef are not statistically different from zero percent.

Table 6. Willingness to pay for organic roast by price variations relative to prices paid for regular beef

	Price variation	Obs. (*)	Responses "Yes"		Responses "Yes, but less"	
		n.	n.	%	n.	%
Price decreases	>10 %	22	22	100.0	0	0.0
	0.01–10 %	18	17	94.4	1	5.6
Bid price equal to price paid	0	26	23	88.5	2	7.7
Price increases	0.01–10 %	21	19	90.5	2	9.5
	10.01–20 %	37	24	64.9	12	32.4
	20.01–30%	16	10	62.5	4	25.0
	30.01–40%	19	4	21.1	14	73.7
	40.01–60%	20	0	0.0	15	75.0
	>60%	22	0	0.0	7	31.8

(*) Data from 201 observations.

Table 7. Willingness to pay for organic minute steak by price variations relative to prices paid for regular beef

	Price variation	Obs. (*)	Responses "Yes"		Responses "Yes, but less"	
	%	n.	n.	%	n.	%
Price decreases	>10 %	21	20	95.2	1	4.8
	0.01–10%	17	17	100.0	0	0.0
Bid price equal to price paid	0	23	20	87.0	2	8.7
Price increases	0.01–10%	23	18	78.3	3	13.0
	10.01–20%	41	27	65.9	11	26.8
	20.01–30%	17	4	23.5	8	47.1
	30.01–40%	33	6	18.2	17	51.5
	40.01–60%	34	7	20.6	18	52.9
	>60%	19	1	5.3	9	47.4

(*) Data from 228 observations.

there is and the more partial substitution of regular quality for organic beef (as shown by the increase of the share of "yes, but less" responses). If prices are too high, the partial substitution is not adopted and the share of "yes, but less" responses declines.

3.3. Average maximum prices consumers are prepared to pay for organic beef

The main point of interest the research focused on was the maximum price consumers are willing to pay for organic beef, since this information sets a higher bound for prospective producers.

In the way the question was asked in the questionnaire, one can get the information whether the maximum price is above or below the proposed price. Different methods can be employed to estimate from this information the average maximum price. One is a non-parametric method, similar in its concept to the calculation of the minimum legal willingness to pay (Harrison and Kriström, 1995). This method exploits the double-ended format of the responses: respondents were asked their willingness to buy with two following bid prices, contingent on their first response. It is then assumed that their maximum willingness to pay is the median of the price interval at which the response is located. For instance, if the respondent answered "yes" to a bid price of 25,000 ITL/kg., and "no" to the second bid price of 30,000 ITL/kg, his/her maximum price is assumed to be 27,500 ITL/kg. Of course, for those who responded "yes-yes" and "no-no", the only available information is that their maximum price is above and below, respectively, the second bid price. Absolute highest and lowest prices have therefore been assumed for the calculation. Lowest prices were assumed to be 10,000 and 12,000 ITL/kg for roast and minute steak, respectively, since it seems very unlikely that lower prices could be found, while several highest prices were experimented.

Given the format of the responses, one may estimate the maximum price both under the hypothesis that consumers totally subsitute organic for regular beef and purchase the same quantity, and under the more realistic hypothesis that they can combine both qualities and adjust the relevant quantities. In the first case, only the "yes" responses have to be considered as positive. The results indicate the maximum price at which consumers are prepared to buy the same quantity of organic as they did of regular beef. With the second hypothesis, the "yes, but less" responses are also considered positive. These results show the maximum price at which a minimum quantity of organic beef is purchased. The results are shown in Tables 8 and 9.

The unconstrained maximum prices range from 1.5 to 2 times the average price of regular beef. The constrained ones are 30 to 45 percent higher than the average price of regular beef.

These results are obviously contingent on the assumed absolute highest prices and it would be desirable to have more precise measures. An alternative way of assessing the maximum price consumers are willing to pay is through a parametric approach. The theoretical and econometric background is as follows (for more details, see Corsi and Novelli, 2003). Assume a consumer has made his/her choice when organic beef was not available. The expenditure function indicates the minimum expenditure needed to

Table 8. Average maximum price for organic roast (ITL/kg)—non-parametric method

Assumed absolute highest prices	Some organic beef	Same quantity as regular
50,000	38,872	31,991
60,000	43,090	33,761
70,000	47,308	35,531
80,000	51,527	37,301

Absolute minimum price has been assumed to be 10,000 ITL/kg.

Table 9. Average maximum price for organic minute steak (ITL/kg)—non-parametric method

Assumed absolute highest prices	Some organic beef	Same quantity as regular
50,000	40,890	33,744
60,000	44,849	35,022
70,000	48,807	36,300
80,000	52,765	37,578

Absolute minimum price has been assumed to be 12,000 ITL/kg.

achieve his/her utility. When organic beef becomes available, if the expenditure now needed for reaching the same utility is less, the consumer will purchase some organic beef. While, if the opposite holds, he/she will consume no organic beef. In other words, the condition for a positive consumption of organic beef is that the difference between the expenditure needed for achieving the same utility, before and after organic beef is available (DE), is positive. If the DE is expressed as a function of explanatory variables and of a random term, then assuming an appropriate distribution for the random term allows estimation of the DE function with maximum likelihood methods. Explanatory variables include the prices for regular and organic beef, income and taste shifters such as socio-economic characteristics. Of course, the higher the price of organic beef the lower is DE, ceteris paribus. The maximum price the consumer is willing to pay for organic beef is the one corresponding to the DE being equal to zero. Therefore, setting the estimated DE equation to zero and solving for the price of organic beef yields a reservation price (RP) equation. The equation gives the maximum price consumers are willing to pay as a function of the explanatory variables. The reservation price for each consumer in the sample can then be calculated by multiplying the individual covariates by the vector of the coefficients of the RP equation. The sample mean and other descriptive statistics for the sample can be used to estimate the relevant parameters in the population. Confidence intervals for the estimates can also be estimated by simulation methods (Krinsky and Robb, 1986). Conceptually, the average maximum price estimated in this way corresponds to the one labelled as "some organic beef" in the previous tables. A similar methodology is employed for estimating the maximum price consumers are willing to pay when they are given as the only choice to totally substitute organic for regular beef and to purchase the same quantity, which corresponds to the "same as regular" item in the previous tables. The estimation of both the unconstrained and quantity-constrained maximum prices is again made possible by exploiting the two different answers "Yes, I would buy it in the same quantity I'm currently consuming" and "Yes, but I would buy less than what I'm currently consuming".

Table 10 shows the results of this exercise. For estimation, consumers were divided into two groups. The first one included those respondents who consumed regular beef and could indicate the price they paid for it. The second group included those who either did not consume regular beef or did consume it, but could not remember its price.

It can be seen that the results are to a large extent consistent with the previous ones[5]. Those presented in Tables 8 and 9 are contingent upon the hypotheses assumed for the largest absolute price, while those presented in Table 10 rely on the assumption of the density distribution of the error term (a normal distribution, censored at the reservation price). The latter exercise suggest a fairly large variation in the individual unconstrained reservation prices, which is consistent with the existing large variation in prices paid for regular beef. The variation is smaller for the constrained reservation prices.

---

[5] The confidence in the results of the parametric approach is strengthened by considerations concerning the consistency of the estimated parameters of the DE equation with economic theory. For instance, the bid price parameter is positive and significant, which implies that consumers paying higher prices for regular beef are also willing to pay higher prices for organic beef. The income parameter is significant and positive for Group 1, and wealthier consumers are therefore willing to pay higher prices. By contrast, it is not significant in Group 2, which seems consistent with the fact that persons who do not remember the price they paid are included in it, along with people concerned with BSE. This could make them much interested in organic beef regardless of their income.

Table 10. Average maximum price for organic beef (ITL/kg)—parametric method

		Some organic beef			Same quantity as regular		
	Mean	95% confidence interval		Mean	95% confidence interval	
		Lower bound	Upper bound		Lower bound	Upper bound
Roast	N = 376			N = 337		
Group 1	40,842	38,272	45,077	31,199	30,422	32,016
Group 2	49,681	43,124	61,457	32,405	31,029	33,885
Total	*45,261*	*38,584*	*58,527*	*31,802*	*30,526*	*33,616*
Minute steak	N = 397			N = 358		
Group 1	45,116	42,924	48,212	35,186	34,416	35,975
Group 2	45,013	41,832	49,632	34,495	32,970	35,896
Total	*45,064*	*42,212*	*49,015*	*34,841*	*33,255*	*35,952*

Of course, if consumers are free to adjust the purchase quantity and possibly to buy partly regular and partly organic beef (which is more realistic) the maximum price at which they buy some quantity of organic beef is larger than the maximum price at which they would totally substitute organic for regular beef and purchase the same quantity as before. The average unconstrained reservation price for organic roast is 75 percent higher than the average current price for regular roast. The corresponding value for the more expensive minute steak is 53 percent. If the average constrained reservation prices are compared to the current average prices, they are 25 and 20 percent higher for roast and minute steak, respectively.

4. CONSUMERS' MOTIVATIONS FOR NOT PURCHASING ORGANIC BEEF

Those respondents who stated they would not purchase organic beef[6] both at the first and at the second lower bid price were asked about the reasons for their response. Three answers were prompted (the item "other reasons" was never chosen): "The price is too high"; "I wouldn't trust it is really organic"; "Organic beef is not superior to regular beef". The answers (Table 11) are of interest for assessing the impediments to consumption.

The first reason, expressing an economic motivation, is by large the dominant one and it covers almost 15 and 23 percent of the total sample for roast and minute steak. This might not be a surprise to many, but some ambiguity is introduced by the fact that a non-negligible share of those who considered the price for organic beef as too high could not indicate the price they paid for regular beef. Precisely, the shares are 30.9 percent and 35.6 percent for roast and minute steak[7]. This, nevertheless, likely indicates that the price is considered too high in absolute terms, regardless of the usual price of regular beef. This impression is reinforced by the fact that the share of "no-no" respondents are larger for higher prices.

---

[6] More exactly, those respondents consuming regular beef and not answering "Yes, I would buy it in the same quantity I'm currently consuming", and those who did not consume regular beef and stated they would not come back and purchase organic beef at the proposed price.

[7] The shares are calculated over the total of those who were asked the price of regular beef. Those who did not consume it were obviously not asked the price.

Table 11. Motivations of "no-no" responses

Motivations	Roast			Minute steak		
	Number of "no-no" responses	% over total no. of "no-no" responses	% over total number of respondents	Number of "no-no" responses	% over total number of "no-no" responses	% over total number of respondents
Price too high	56	70.9	14.8	91	79.8	22.8
I wouldn't trust it is really organic	8	10.1	2.1	8	7.0	2.0
Organic beef isn't superior to regular beef	15	19.0	4.0	15	13.2	3.8
Total	79	100.0	20.8	114	100.0	28.6

Although high price is the main impediment for refusing organic beef, the other responses should not be disregarded since they show that trust is an important determinant of consumption. About 6 percent of the sample stated that the reason for not consuming organic beef is either skepticism in the real organic quality or distrust about the control over organic quality. Here there is room for an action by organic producers' organizations and by public bodies to inform consumers about the differences between regular and organic produce and about the control system.

5. PREFERENCES ABOUT SELLING MODALITIES

Another point of interest for prospective producers was selling modalities. Respondents were asked about their preferred modalities. In particular they were asked whether they would be more trustful when purchasing packaged beef with a label indicating all required data or beef cut by a butcher in a selling point showing a certification document of the animal. Those who preferred the latter (plus the "indifferent" and "don't know") were further asked their degree of trust in butchers (other than their usual one, to avoid personal trust) selling both regular and organic beef (*mixed*), and in butchers only selling organic beef (*exclusive*).

The large majority of respondents (61.2 percent) preferred organic beef directly cut by butchers, as compared to 19.9 percent who indicated packaged beef. The rest was indifferent (15.9 percent) or did not know (3 percent). The preferences are to a large extent related to purchasing habits. Eighty-five point three percent of those usually buying beef at a butcher's would prefer to purchase organic beef in the same way and 64 percent of those usually shopping in supermarkets indicated packaged organic beef. As to the degree of trust towards *mixed* and *exclusive* butchers, there is a large majority of "fairly" responses for both, but *exclusive* butchers are clearly preferred (Table 12)[8].

---

[8] A chi-square test strongly rejects the hypothesis of independence of the responses from the type of butcher.

Table 12. Degree of trust towards *mixed* and *exclusive* butchers (%)

Would trust:	Mixed butcher	Exclusive butcher
not at all	31.1	14.3
fairly	61.5	68.3
much	3.4	14.0
don't know	4.0	3.4
Total	100.0	100.0

6. CONCLUSIONS

Overall, the results of all exercises suggest that organic beef could easily find an appreciable market share and that there are consumers prepared to pay substantially higher prices than the ones they are paying for regular beef. This might be an encouraging signal for prospective producers of organic beef, since the likely increase in production costs might be compensated for by a substantial premium for the new quality. Of course, this conclusion relies on the consistency of actual purchasing behavior with stated preferences. Nevertheless, there are no particular reasons why an inquiry about organic beef should be biased. The results are clear enough to justify an optimistic view of market prospects.

This conclusion seems at odds with the fact that, at the moment, consumption of organic beef is still very limited in Italy and the number of organic beef producers is very scarce. At the end of 2002, there were less than 3,000 organic meat producers (INEA, 2003). The slow growth of the organic beef sector seems to be mainly due to difficulties on the supply side. Some of these reasons include the lack of clarity in the European Regulation, specifically concerning pasture management, and technical and economic problems in complying with the Regulation. Most regular beef cattle is reared in intensive farms in Italy, largely based on purchased animal feed. The high price of organic maize (the most used animal feed) would impose high costs on organic beef producers. Organic producers have to be self-sufficient for a minimum share of animal feed and high production cost are implied for self-produced maize. Producers of native breeds, usually reared in marginal areas and largely relying on pasture, are a category that could more easily comply with the organic production requirements. In their case, though, prices of regular beef are already high and further price increases for organic quality could hardly find a market. Moreover, there are some technical problems with the finishing period when following organic rules. Those producers based on summer grazing on the Alps have some problems too. In their case, they would have to meet the animal welfare requirements of winter cattle housing systems that is often not adequate and should be enlarged to provide the required space for animals. Other reasons for the weak consumption of organic beef might be that supermarkets are usually reluctant to engage in selling a new product unless sufficient supply in regular times is guaranteed. Organic beef is therefore mainly sold in specialized butchers. In this way only committed consumers actively looking for the product purchase it, occasional consumers are not induced to consumption. Finally, the present sluggish economic situation is not favorable to a more expensive product.

7. REFERENCES

Andreoni, J., 1990, Impure altruism and donations to public goods: a theory of Warm-glow giving, *The Economic Journal*, **100**: 464–477.

Carson, R.T., Hanemann, W.M., and Mitchell, R.C., 1986, *Determining the demand for public goods by simulating referendum at different tax prices*, Manuscript, University of California, San Diego.

Corsi, A., and Novelli, S., 2003, *Measuring Quantity-Constrained and Maximum Prices Consumers are Willing to Pay for Quality Improvements: the Case of Organic Beef Meat*, paper selected for presentation at the 25th International Conference of Agricultural Economists, August 16–22, 2003, Durban, South Africa (CD-Rom).

Kahneman, D., and Knetsch, J.L., 1991, Valuing public goods: the purchase of moral satisfaction, *Journal of Environmental Economics and Management*, **22**: 57–70.

Krinsky, I., and Robb, A.L., 1986, On approximating the statistical properties of elasticities, *Review of Economics and Statistics*, **68**: 715–719.

Hanemann, W.M., Loomis, J.B., and Kanninen, B.J., 1991, Statistical efficiency of double-bounded dichotomous choice contingent valuation, *American Journal of Agricultural Economics*, **73**: 1255–1263.

Harrison, G.W., and Kriström, B., 1995, On the Interpretation of Responses to Contingent Valuation Surveys, in P.O. Johansson, B. Kriström, and K.G. Mäler (eds.), *Current Issues in Environmental Economics*, Manchester University Press, Manchester, pp. 35–57.

INEA, 2003, *Annuario dell'Agricoltura Italiana*, Volume LVI, 2002, ESI, Napoli.

Marengo, G., Bassignana, E., Corsi, A., and Didero, L., 2000, *Le prospettive del mercato dei prodotti zootecnici da agricoltura biologica*, Regione Piemonte—Assessorato Agricoltura, Torino.

THE US CONSUMER PERSPECTIVE ON ORGANIC FOODS

Carolyn Dimitri and Luanne Lohr[*]

SUMMARY

Market growth in the US organic sector has been dramatic since the mid-1990s. Much of this growth was made possible by rapidly increasing consumer interest, which fostered new market opportunities for organic food producers, retailers, and distributors. Market distribution networks have changed character as organic and conventional firms have joined forces, catapulting organic foods into mainstream grocery stores within the reach of most consumers. Consumers now buy organic food for health, taste, and social factors, in additional to the traditional environmental reasons. Even as consumers buy organic food for a wider variety of factors, direct sales from farmers to consumers have remained an important component of the organic market as the market has grown.

1. INTRODUCTION

Unarguably, today's US organic food sector is dynamic. Growing consumer demand culminated in the near tripling of organic retail sales between 1997 and 2003, from USD 3.6 million to USD 10.4 million, according to industry estimates. Much of this growth was made possible by rapidly increasing consumer interest, which fostered new market opportunities for organic food producers, retailers, and distributors. As the sector has grown, a new picture of consumers and their reasons for purchasing organic food has emerged. Simultaneously, market distribution networks have changed character as organic and conventional firms have joined forces, catapulting organic foods into mainstream grocery stores within the reach of most consumers.

Part of what makes the organic market unique is that the businesses and consumers it serves judge food not only by its taste, price and appearance, but also by the social and

---

[*] Carolyn Dimitri, Economic Research Service, USDA, Washington DC, 20036. Luanne Lohr, University of Georgia, Athens, 30602.

environmental benefits it represents. In this respect, it is a market that does not rely solely on economic factors in defining its products. For example, consumers may purchase organic foods for social reasons, such as buying locally grown organic food direct from farmers to support local farmers or buying fair trade organic coffee or chocolate to support fair wages for farmers, while others may buy organically grown food for environmental reasons. These core organic consumers have been joined by a new group, consisting of those who purchase organic food for health or taste.

Consumers report strong preferences for locally grown foods that are available primarily through direct-marketed outlets. Numerous surveys of farmers' market consumers, Community-supported agriculture (CSA) subscribers, and farm stand buyers have delineated several reasons for the preference for locally grown foods. Locally grown foods are perceived to be fresher and taste better. The ability to speak face-to-face with the grower and possibly visit the farm where the food was grown instills confidence in the environmental and human safety of the production method. Consumers report a desire to support the local farm economy, protect agricultural green space, and the local food system.

This chapter provides an overview of the US organic market for organic products, from the perspective of the consumer. We first detail trends in retail sales and farmer outlets for organic products, and next discuss the reasons consumers purchase organic food. We then shift focus to direct sales of organic foods; this in-depth discussion reflects the importance of farmers' markets and other outlets for both consumers and producers of organic foods.

2. US MARKET FOR ORGANIC FOOD PRODUCTS

USDA does not have national statistics on organic retail sales. Industry sources have reported retail sales for organic food, but those data are fragmentary and, at times, inconsistent. A trade publication, The *Natural Food Merchandiser* (NFM) reported estimates of total U.S. retail sales of organic foods for 1990 as USD 1 billion and in 1996 as USD 3.3 billion. More recent industry estimates were reported by the *Nutrition Business Journal* (NBJ), which estimates sales of USD 3.47 billion in 1997 to nearly USD 10.4 billion in 2003, or about 2 percent of the country's total retail food bill. Growth rates have remained around 20 percent annually. Estimates through 2004 place growth rates at approximately 10–18 percent, slowing in the later years, with sales of USD 23.78 billion in 2010 (Figure 1), and organic sales representing 3.5 percent of the total retail food bill (NBJ, 2003).

Of the total retail sales for 2003, USD 4.34 billion was from organic fruits and vegetables, representing 42 percent of the total sales in organic foods; fresh produce has long been the top selling organic food category. In 2003, unpackaged produce accounted for USD 1.75 billion in sales, including USD 400 million through farmers' markets of which 95% were estimated to be fresh fruits and vegetables. In a nationwide consumer survey, the Hartman Group (2000) found that 14% of retail organic purchases were through farmers' markets, compared with 60% for conventional supermarkets, mass merchandisers, and club stores. The consumer survey found that the majority of organic purchases from farmers' markets were fresh fruits and vegetables.

In 2003, retail sales of organic beverage totaled USD 1.58 billion, followed by dairy (USD 1.39 billion), packaged/prepared food (USD 1.33 billion), breads and grains (USD

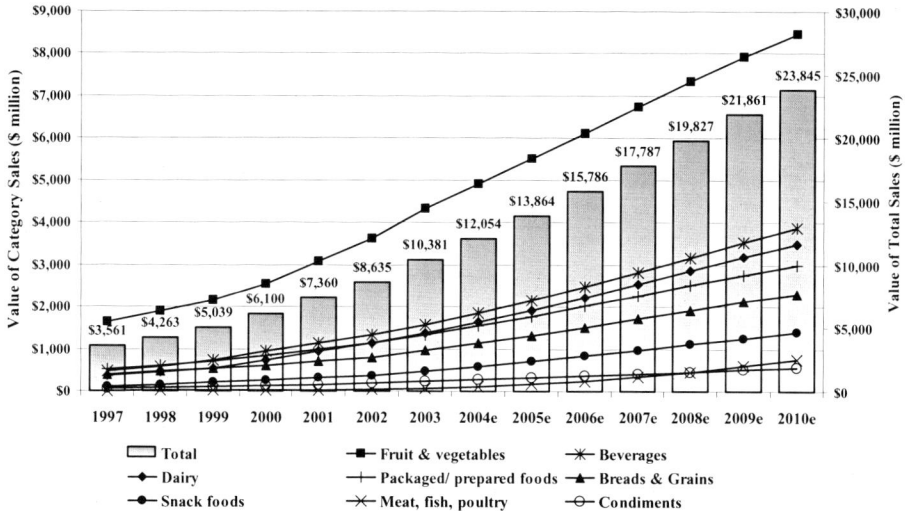

Figure 1. Value of Category and Total Sales in Organic Food Industry, 1997–2010, estimated for 2004 to 2010 (Source: Nutrition Business Journal, 2004)

966 million), snack foods (USD 484 million), condiments (USD 229 million), and meat, fish, and poultry (USD 75 million) (NBJ, 2004). Sales in each of these categories, as Figure 1 shows, have been steadily increasing since 1997.

Growth in retail sales of milk, half and half, and cream were second to that of fresh cut produce, measuring 21 percent between 2002 and 2003, and nearly three-fourths of the USD 304 million was sold in conventional supermarkets. Growth rates understate increases in demand because the dairy sector has been plagued by milk shortages, despite supply increases resulting from larger herds of the major milk manufacturers as well as entry by new private label dairies (Blank, 2004). Some speculate that increased demand for organic milk products is the result of exceptionally high conventional milk prices in 2002; organic milk prices did not rise, so consumers responded to the declining organic premium by substituting organic milk for conventional milk.

A growing trend has been for longtime manufacturers of conventional products to introduce organic items to their product lines, typically by acquiring smaller organic firms. The conventional manufacturers have made use of their extension distribution networks to place organic products in conventional retail outlets, and as a result, organic food is sold in a wide variety of retail outlets. In 2003, 43 percent of all organic food products were sold in mainstream channels, such as supermarkets, drug stores, and mass merchandisers, with 47 percent sold in natural product supermarkets. In comparison, in 1991, 7 percent of organic retail sales occurred sold in mainstream channels in 1991 and 68 percent in natural product stores (Table 1).

According to the *Natural Foods Merchandiser,* of 10 categories, only three (yogurt & kefir, soups, and bread & baked goods) had larger sales in natural product supermarkets. Sales of breakfast cereals, in contrast, exploded in 2003, increasing by 46 percent;

Table 1. Distribution of organic food sales by channel in 2003 and 1991.

Channel	Sales (millions USD)	Share of organic sales, 2003	Share of organic sales, 1991
Conventional Grocery	3,868	0.37	.07
Natural Foods Grocery Chain	2,011	0.19	.68
Natural Foods Independent Store	2,932	0.28	
Mass Merchandiser	367	0.04	
Club Store	264	0.03	
Other	120	0.01	.25
Food Service	254	0.02	
Export	165	0.02	
Farmers' Market	400	0.04	

Note: in 1991, farmers' market is included in the "other" category.
Source: OTA's 2004 Manufacturer Survey; Dimitri and Greene, 2002.

One significant factor leading to this increase was the impact of the acquisition of two organic cereal firms by large conventional cereal manufacturers that were able to manufacture and distribute organic breakfast cereals to mainstream supermarkets nationwide. Mainstream channels also sell the lion's share of packaged fresh produce, at 75 percent; again, likely made possible by joint ventures between organic brands and conventional fresh cut produce firms.

Tracing sales from the farm-level indicates a similar shift in outlets available to farmers, as Table 2 documents. One of the most marked changes is the decline in the share of organic food sold in the consumer-direct outlet. As market mature and more and larger organic farms are certified, the means of selling farm output is changing. Consumer-direct outlets offer few barriers to entry for low volume producers and typically do not incur significant investment costs to use (Georgia Organics, 2004). As farms get larger, producers are able to deliver sufficient quantities of output to secure retail and wholesale contracts. The guaranteed return of a wholesale or retail contract is a strong stimulus for a farmer to shift all or part of output into retail or wholesale market channels.

For fruits, nuts, and tree fruits, volume was reallocated from consumer-direct and direct to retail sales into wholesale outlets, which grew from 51% to 77% between 1997 and 2002. Sales to handler/brokers and processors accounted for most of the gain, with both outlets nearly doubling their share of volume handled from farms. In grains and field crops, there consumer-direct sales volume held steady over the five year period. Livestock and animal products was the only category to increase consumer-direct sales, increasing mainly due to Internet and mail order sales, which rose from 1.5% of volume to 5.0% between 1997 and 2002 and to subscription sales, which grew from 0.8% to 1.7%. For vegetables, the decline in consumer-direct share of farm output volume is due to gains in direct to retail sales, which accounted for 53% of volume in 2002 compared with 19% in 1997. Direct to retail sales of fresh vegetables are common practice for conventional supermarkets (Kaufman et al., 2000), which experienced a fourfold increase in share of volume distribution of organic vegetables.

Table 2. Percent of organic farm output volume by marketing channel and product category, 1997 and 2002[a]

Marketing channel	Vegetable		Fruit		Field Crop		Livestock	
	1997	2002	1997	2002	1997	2002	1997	2002
Consumer-direct	22.8	12.9	27.9	11.0	11.4	10.9	20.3	25.5
– Direct on-farm	9.4	2.5	18.5	4.7	7.5	6.6	15.3	4.5
– Farmers' market	7.7	5.6	5.8	3.9	1.7	0.1	2.7	3.3
– CSA or subscription	3.6	4.5	2.4	1.0	0.6	0.4	0.8	2.7
– Other consumer-direct	2.1	0.3	1.2	1.4	1.6	3.8	1.5	5.0
Direct to retail	18.9	53.2	21.6	12.2	6.4	1.1	7.9	12.2
Wholesale	62.0	33.9	50.9	77.2	82.3	88.2	71.8	62.3

[a] The vegetable category includes vegetables, herbs, floriculture, mushrooms, and honey. The fruit category includes fruit, nut, and tree products. The field crop category includes grain and field crops. The livestock category includes livestock and animal products.
Sources: Walz, 1999 and 2004.

3. THE US ORGANIC FOOD CONSUMER

The portrait of the typical organic food consumer has changed over time, reflecting the dynamic nature of the organic industry. Recently (in 2002), most studies characterized organic consumers as Caucasian, affluent, well-educated, and concerned about health and product quality (Lohr, 2001;Richter et al., 2000; ITC, 1999; Thompson, 1998). While this type of consumer still purchases organic food, current consumers of organic food are far more diverse and not as easily characterized. Income and ethnicity are no longer significant predictors of who purchases organic food: Half of those who frequently buy organic food have incomes below 50,000, and people of color—African Americans, Asian and Americans and Hispanics—use more organic products than the general population does (Howie, 2004). In 2004, 42 percent of organic consumers had incomes below USD 40,000 (Barry, 2004). The average age of organic consumers is clustered in two age groups: 18 to 29 years and the 40s (Thomson, 1998; Lohr and Semali, 2000). One element has remained constant as the industry grows, that being many consumers are parents of young children or infants.

Consumer buying patterns follow a fairly predictable path. The first organic products consumers typically buy are produce, dairy products, nondairy products (such as soy milk), and baby food. The second group of products usually consists of juice, single serving beverages, meat/poultry, cold cereal and snacks. Next, consumers expand their purchases to include frozen foods, breads, pasta sauces, salsas, and canned tomatoes. Other organic canned goods and bulk goods are usually next purchased, with non food products (fibers) the last organic items consumers try (Demerrit, 2004).

Again, reflecting the diverse nature of the market, the organic consumer of 2004 has many reasons for buying organic food. Consumer surveys indicate that health concerns have emerged as the primary motivation for purchasing organic food. Consumers perceive that, by purchasing organic foods, they can positively affect their health by avoiding pesticides, antibiotics and growth hormones, and GMOs (Hartman Group, 2000). Other consumers favor environmentally friendly products, and organic food is consistent with this preference. Taste and food safety are the next two most often given reasons for purchasing organic food. Age, gender, and having a college degree had little impact on a shopper's decision to buy organic produce (Thompson and Kidwell, 1998). Appearance of fresh produce mattered, and the larger the number of cosmetic defects, the less likely would an organic product be purchased (Estes and Smith, 1996; Thompson and Kidwell, 1998).

Organic foods are typically more expensive, costing 10–30 percent more than their conventional counterparts (Lohr, 2001). Surveys indicate mixed results about consumer response to higher priced organic food. A 2003 survey indicates 73 percent of consumers believe organic food is too expensive (Whole Foods Market, 2003), confirming other studies indicating that price was a barrier (The Packer, various years; Walnut Acres, 2001). Higher prices appear to be less of a barrier for those organic products highly valued by consumers, such as fresh produce or baby food (Barry, 2004). Econometric analysis indicates that organic and conventional milk and baby food are substitutes, so that increases in the price of the conventional product result in consumers' purchasing a greater quantity of the organic products (Glaser and Thompson, 2000; Thompson and Glaser, 2001).

Consumers who want to advance social goals such as equitable income distribution and sustainable development have the option of purchasing goods certified as "Fair Trade". Speaking generally, Fair Trade certification standards require that traders pay producers prices that cover costs of sustainable production and living; pay producers a premium so they can invest in development; pay producers in advance (partially) when requested; and enter long term contracts with producers that make it possible for producers to make long term plans and adopt sustainable production practices (TransFair USA, 2005). Fair Trade certification differs from organic certification, although 65 to 85 percent of Fair Trade imports also carry organic certification (ITC, 1999). Of the USD 180 million of fair trade products sold in North America in 2002, USD 131 million of sales were for fair trade coffee, an increase of 54 percent from 2001. One factor contributing to the huge increase was Starbucks' introduction of Fair Trade coffee in 2001, which made fair trade coffee widely available. Fair trade cocoa is another agricultural product with fast growth between 2001 and 2002.

4. DIRECT MARKET SALES OF ORGANIC FOOD PRODUCTS

Where and how food is produced matters to a significant portion of organic consumers. This local preference incorporates ethical views toward farming and local growers. Interest in supporting regional producers is strong among regular buyers of organic foods (Richter *et al.*, 2000). As a result, there is enormous growth potential for consumer-direct marketing, which includes farm-direct sales, farmers' markets, subscription farming, mail order, internet sales, and other marketing outlets. Farm-direct outlets include farm stands, roadside stands, u-pick operations that offer a single farmer's outputs, while farmers' markets offer products from multiple farmers meeting in a single location on specified days of the week. Subscription farming includes Community Supported Agriculture, in which subscribers pay an up-front fee in exchange for a weekly share of farm products provided throughout the growing season. Mail order sales using catalogs are fast being updated by Internet sales that use a web page to display product images and descriptions and permit online ordering. All other direct marketing methods may include barter exchanges, sales to other farmers, and sales at fairs and festivals.

Consumers and producers both may gain from consumer-direct sales. With the exception of mail order and Internet sales, consumer-direct sales usually take place within 100 miles of the farm location, providing consumers the opportunity to obtain fresher products, and even to physically visit the farm on which the food was produced.

Frequently, locally grown organic products purchased directly from farmers are available at a lower price than those sold through retail outlets, with cost savings of the shorter supply chain offsetting the cost advantages of retailer buying power in stocking organic foods. The consumer-direct cost savings are realized in part through energy cost savings for transportation and storage as distance between origin and consumer increases, and are magnified with increasing oil prices.

Consumer-direct marketing especially benefits farmers producing low volumes of a variety of crops. Benefits to producers of consumer-direct sales include cash payment, premium pricing, and ability to retain a larger share of the food dollar paid by the consumer. In consumer-direct sales, farmers typically must charge more than they would a wholesaler to cover the costs of handling, storing, packing, transporting, and advertising their products in addition to growing them. Although farmers retain more of the consumer food dollar, the difficulty in identifying and reaching consumers may impose a significant burden on the farmer. Most farmers do not obtain sufficient income from a single consumer-direct marketing outlet, and many use a combination of consumer-direct, direct to retail, and wholesale marketing strategies.

Organic farmers have traditionally consumer-direct marketed a larger share of output than conventional farmers. Among all farmers, the percentage of farms marketing direct to consumers rose slightly from 5.2% in 1997 to 5.5% in 2002 (USDA, 2004). Among organic farmers this percentage ranged from 11% to 26% in 2002, depending on product category (Walz, 2004). However, as retail sales of organic foods have increased, the

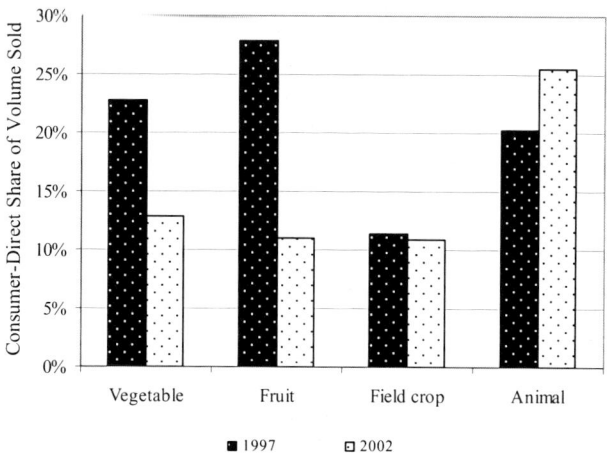

Figure 2. Consumer-direct share of volume sales by organic farms, 1997 and 2002 (Source: Walz, 1999 and 2004)

consumer-direct share of organic farm sales volume has declined for most product categories. Figure 2 shows that more than 20% of fresh fruit and vegetable volume was sold directly to consumers in 1997, dropping to less than 13% by 2002. Organic livestock sales directly to consumers have increased as the category has grown, rising from about 20% to about 25% between 1997 and 2002, much of it as fresh or frozen farm-processed cuts and cured meats.

In 2002, 51% of organic farmers reported an intention to increase consumer-direct sales, while only 4% planned to decrease sales through this channel (Walz, 2004). As the number of organic farms grows, it is safe to assume that entry-level marketing outlets such as farmers' markets and farm stands will continue to be cost-effective for smaller and newer farms.

Although farmers are willing to supply outputs direct to consumers, it is not clear how committed consumers are to this marketing channel. The Hartman Group (2000) reported that heavy buyers of organics, described as purchasing on average 28 organic items in the three months prior to the survey, access the Internet organic sales channel about twice as frequently as light buyers do, but buy about the same percentage, 2% of their total purchases, from Internet sites. Heavy buyers purchase slightly less from farmers' markets than light buyers, approximately 12% of their overall organic purchases compared with 14%.

The trend toward more buyers, but a smaller percentage of regular buyers of organic foods is reflected in category data collected by HealthFocus International (2001). Figure 3 shows the percentage of consumers who buy organics at least twice weekly from 1994 to 2000. For grains, fruits and vegetables, and processed foods such as soups and sauces, the percentages have declined. The only increase in twice-weekly or more frequent purchases is for meats, which tracks with rising concern over bacterial and viral contamination such as *E. coli* and Bovine Spongiform Encephalopathy (BSE). However, in general, a smaller percentage of consumers in 2000 believed that organic foods were safer to eat than in 1994, 59% compared with 75% (HealthFocus International, 2001).

A requirement for consumer-direct sales is the ability to offer products that consumers want to purchase. The Hartman Group (2000) described four organic buying demographics, representing 31% of U.S. consumers who are multi-product organic users. Specialty Foodies make up 29% of the organic buying segment and purchase many stereotypical health foods outside the mainstream shopping experience, such as soy and rice milk, herbal teas, tofu, ethnic meals, and organic soft drinks. Pacific Produce Pickers make up 35% of the segment, and buy primarily fresh fruits and vegetables, coffee, and nuts. The Miss American Pie demographic is 20% of the segment, and purchases products deemed typical of the American experience including condiments, ice cream, meats, sauces, and dairy products. Topper Shoppers are only 16% of the segment, but buy on average 49 different products. These consumers purchase primarily condiments, sauces, and dairy products.

For a typical direct-market operation selling fresh produce, meats, and dairy products, the primary consumer target groups would be the Miss American Pie and Pacific Produce Pickers segments. With value added on-farm processing, the Topper Shoppers and Specialty Foodies offer another segment for consumer-direct marketing. Walz (2004) reported that about 59% of organic farmers derive some income and 11% derive more than 75% of their gross sales from value-added processing. The most common items are fresh or frozen cut meats, herbal teas and supplements, cosmetics, canned or bottled

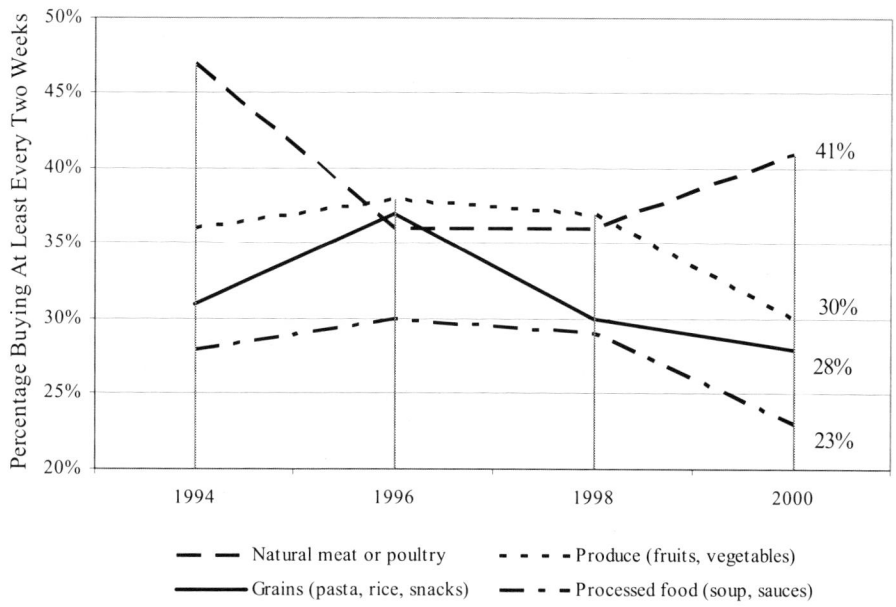

Figure 3. Percentage of consumers reporting organic purchase at least every two weeks, by product category, 1994 to 2000 (Source: HealthFocus International, 2001)

vegetable products, salad mixes, preserved fruits, juices and cider, wine, and dried fruits and vegetables. Most on-farm value-added processing involves few ingredients and low-cost technologies. Nearly 34% of organic producers plan to increase the number of value-added items they market (Walz, 2004).

A trend that began around 2000 and was accelerated by the terror attacks of September 11, 2001 is a heightened interest in local and regional food security, providing quality foods at a reasonable price for all citizens while minimizing the miles food must travel for retail sale. Farmers' markets have expanded services to families served by the USDA Women, Infants and Children (WIC) program through coupon redemption programs for fresh produce. While the preference for locally grown food is not specific to organic, the organic sector is uniquely able to respond to this demand. With a larger percentage of farmers already engaged in consumer-direct marketing and greater farm diversification, the typical organic farm can offer more local products to consumers.

5. THE FUTURE OF THE US ORGANIC FOOD MARKET

The widely held myth that only high income consumers will buy organic foods is being dispelled. Organic farmers who offer foods in inner city farmers' markets face no greater price resistance than conventional farmers who market in areas of low-income consumers who comparison shop at mass market retailers. Education about organic production methods and experience with product quality can convince low-income buyers that organic food is worth a small price premium, just as it does for high-income consumers. Examples in Chicago, Detroit, and San Antonio converting vacant lots into organic gardens and bringing in local produce for inner city farmers' markets help disprove the myth. Surveys indicate that a diverse group of consumers is currently buying organic food, crossing ethnic, economic, and age groups.

The outlook for organic agriculture in the US is bright. Expansion of consumer-direct organic markets may take many directions, from Internet sales of specialty or locale-specific items to farmers' markets in low-income urban and rural areas to subscription farming enterprises on the outskirts of small towns. Offerings of organic food in traditional retail outlets, such as natural products stores and conventional grocery stores, appear poised to grow over the next decades. If, as The Hartman Group (2000) reported, 60% of American consumers do not eat organic foods but are willing to try them, the potential for growth is enormous.

6. REFERENCES

Barry, M., 2004, The New World Order: Organic Consumer Lifestyle Segmentation, *[n]sight Magazine*, **VI**(2): 3–7.
Blank, Christine, 2004, High Demand for Dairy Products Creates Shortage for Organic Suppliers. *Organic Business News*. December 2004. **16**(12): 6–7.
Demeritt, L., 2004, Organic Pathways, *[n]sight Magazine*, **VI**(2): 16–21.
Dimitri, C., and Greene, C., 2002, *Recent Growth Patterns in U.S. Organic Foods Market*, Agricultural Information Bulletin No. 777. U.S. Department of Agriculture, Economic Research Service.
Estes, E., and Smith, V.K., 1996, Price, Quality, and Pesticide-Related Health Risk Considerations in Fruit and Vegetable Purchases: An Hedonic Analysis of Tucson, Arizona Supermarkets, *Journal of Food Distribution Research*, **27**(3): 8–17.
Georgia Organics, 2004, *Choosing Your Market: A Direct Marketing Decision Tool*, Georgia Organics, Inc., Atlanta, GA.
Glaser, L., and Thompson, G., 2000, *Demand for Organic and Conventional Beverage Milk*. Paper presented at the Western Agricultural Economics Association Annual Meetings. Vancouver, British Columbia. June 29-July 1.
Hartman Group, 2000, *The Organic Consumer Profile*, The Hartman Group, Bellevue, WA.
Hartman Group, 2002. Hartman Organic Research Review: A Compilation of National Organic Research Conducted by the Hartman Group. Bellevue, Washington.
Howie, M., 2004, Research Roots out Myths Behind Buying Organic Foods, *Feedstuffs*, March 29.
HealthFocus International, 2001, *What Do Consumers Want From Organics?*, HealthFocus International, St. Petersburg, FL.
International Trade Centre (ITC), 1999, *Organic Food and Beverages: World Supply and Major European Markets*, ITC/UNCTAD/WTO, Geneva.
Kaufman, P., Handy, C.R., McLaughlin, E.W., Park, K., and Green. G.M., 2000, *Understanding the Dynamics of Produce Markets: Consumption and Consolidation Grow*, Agriculture Information Bulletin No. (AIB758), ERS-USDA.
Lohr, L., 2001, Factors Affecting International Demand And Trade in Organic Food Products, in: *Changing Structure of Global Food Consumption and Trade*, A. Regmi, ed., WRS No. (WRS01-1), ERS-USDA, Washington, DC, pp. 67–79.
Lohr, L., and Semali, A., 2000, Retailer Decision Making in Organic Produce Marketing, in: *Integrated View of Fruit and Vegetable Quality*, W.J. Florkowski, S.E. Prussia, and R.L. Shewfelt, eds.,, Technomic Pub. Co., Inc., Lancaster, PA, pp. 201–208.
Natural Foods Merchandiser, various issues, New Hope Natural Media, Boulder, CO.
Nutrition Business Journal (NBJ), 2003. *The NBJ/SPINS Organic Foods Report 2003*. Penton Media, Inc.

Nutrition Business Journal (NBJ), 2004, *U.S. Organic Food Sales, 1997–2010e—Chart 22*, electronic spreadsheet available from publisher, Penton Media, Inc., San Diego, CA.
Organic Trade Association (OTA), 2004, *Organic Trade Association's 2004 Manufacturer Survey*, Organic Trade Association, Greenfield, MA.
Richter, T., Schmid, O., Freyer, B., Halpin, D., and Vetter, R., 2000, Organic Consumer in Supermarkets—New Consumer Group with Different Buying Behavior and Demands!, in: *Proceedings 13th IFOAM Scientific Conference*, T. Alfödi, W. Lockeretz, U. Niggli, eds., vdf Hochschulverlag AG and der ETH Zürich, pp. 542–545.
The Packer, 2002, 2000, 1998, 1996, *Fresh Trends: Profile of the Fresh Produce Consumer*, Vance Publishing, Shawnee Mission, KS.
Thompson, G.D., 1998, Consumer Demand for Organic Foods: What We Know and What We Need to Know, *American Journal of Agricultural Economics*, **80**: 1113–1118.
Thompson, G.D., and Glaser, L., 2001, *National Demand for Organic and Conventional Baby Foods*, Selected Paper presented at Annual Meeting of the Western Agricultural Economics Association, Logan, Utah, July 8–11.
Thompson, G.D., and Kidwell, J., 1998, Explaining the Choice of Organic Produce: Cosmetic Defects, Prices, and Consumer Preferences, *American Journal of Agricultural Economics*, **80**(2): 277–287.
TransFair USA, 2005; http://www.transfairusa.org.
U.S. Department of Agriculture (USDA), 2004, *2002 Census of Agriculture—United States Summary and State Data*, Volume 1, Geographic Area Series, Part 51, AC-02-A-51, Table 2. Market Value of Agricultural Products Sold Including Landlord's Share, Direct, and Organic: 2002 and 1997, National Agricultural Statistics Service, USDA, Washington, DC.
Walnut Acres survey, 2001, Boulder, CO; http://www.walnutacres.com, available online December 2002.
Walz, E., 1999, *Final Results of the Third Biennial National Organic Farmers' Survey*, Organic Farming Research Foundation, Santa Cruz, CA.
Walz, E., 2004, *Final Results of the Fourth National Organic Farmers' Survey: Sustaining Organic Farms in a Changing Organic Marketplace*, Organic Farming Research Foundation, Santa Cruz, CA.
Whole Foods Market, 2003, *One Year after USDA Organic Standards are Enacted More Americans are Consuming Organic Food*, Whole Foods Market, Inc., Austin, TX; http://www.wholefoods.com/company/pr_10-14-03.html.

RECENT DEVELOPMENTS AND FUTURE ISSUES

CURRENT ISSUES IN ORGANIC FOOD: ITALY

Maurizio Canavari [*]

SUMMARY

This chapter suggests some relevant topics for further analysis of organic farming and organic food markets in Italy. The discussion regards different aspects, taking into consideration farmer issues, supply chain issues, consumer issues, policy and trade issues, including farmers' income, sustainable development and international trade. All these aspects show important implications with organic agriculture. A final section takes into consideration further emerging issues, linked to globalisation and to the need to foster further research activities in support of the diffusion of organic farming not only in Italy, but also in developing countries.

1. INTRODUCTION

In this book, some of the relevant issues relating to organic food in Italy from both the consumers' and producers' viewpoints have been analyzed. Further issues and obstacles faced both by farmers and consumers in Italy, that are not explicitly addressed in preceding chapters, even if in some cases they are underlying, are briefly discussed in this chapter.

The aim of the chapter is not to fill the gap of in-depth analysis left by the preceding chapters, it is rather a summarising of relevant research topics, that may be of interest and give raise to further research activities on organic food in Italy.

The chapter is organised in four sections, in which the following topics are discussed: farmer issues, food supply chain issues, consumer issues, policy and trade issues. A final section suggests some recently emerging issues.

---

[*] Department of Agricultural Economics and Engineering, Alma Mater Studiorum-University of Bologna, Bologna, Italy. Thanks to Giovanni Galanti, Bioagrico-op, for suggestions and insight in the organic sector, and to Kent Olson, University of Minnesota, for his review of this chapter. All errors and omissions remain the responsibility of the author.

2. FARMER ISSUES

Organic agriculture in Italy is presently experencing a "crisis of growth." On one side, organic agriculture is getting "out of the niche" to become a small segment, facing interesting opportunities on the mass market; on the other side, the production sector and the organisation level of its supply chain are not fully prepared to cope with wider and more demanding and competitive markets.

Despite the recent developments, however, the main issues for agricultural producers are still in regard to some basic questions listed below. These questions are all potential topics for further research on the situation of organic food in Italy.

- Will organic farming be profitable? From the farmer's perspective, the issues of comparable yields, available technical tools, cost and revenues, crop profitability, and the ease of adopting organic farming are still of paramount importance. These issues depend not only on final retail prices but also on the added value and its distribution along the suppy chain. It is not rare that organic agricultural products are sold as organic at the same price as the conventionally grown ones at farm level. This happens mainly due to the lack of an appropriate distribution system in the area where the product is grown. The problem of production cost is then important, but lower cost is not the only reason that producers are willing to switch towards an organic production system.
- Is an organic product or an organic food product able to maintain a better market potential than a conventionally grown product in the future? While there has been strong support of organic production on the supply side in recent years, the demand side was not cared for enough. The structural excess demand for this kind of products was reduced in the last years and price *premia* at farm levels were often reduced too.
- Is being an organic producer preferred by farmers even if profit differentials are not clearly higher? Are there other conditions and motivations that may encourage farmers to opt in and remain in the organic sector? Some organic farmers actually make the choice to convert to organic production even if the local conditions do not assure a clearly higher profitability of single crops. The consideration of farmers' motivations may include personal attitudes of producers, as well as multifunctionality-linked opportunities, that create the conditions for improving the farmers income, integration into the local community and quality of life. Initiatives such as establishing "teaching farms," farm tourism facilities and leisure activies, direct marketing of fresh produce or artisanally processed food at farm's gate may be some examples.

Beside these topics, other concerns, obstacles and motivations should be taken into consideration. These include the risk of contamination from genetically modified (GM) crops (since they are presently not allowed in organic farming), organic input availability, market information, administrative burden of controls and development pressures near urban areas.

Another important aspect is the compatibility of the organic concept with another powerful rural developement and marketing tool, widely used in Italy, i.e. the geographical indications (PGI/PDO) instruments. Compatibility of production rules is not always guaranteed, and the issue of overlapping controls is also of extreme importance.

3. FOOD CHAIN ISSUES

In this section some concerns, issues, and obstacles related to processing, distribution, and marketing of organic food are briefly discussed.

In this book, the Italian food chain has not been taken into consideration. Actually, there are few economic studies considering food chain issues for organic products in Italy.

Food quality is a concept of crucial importance for the understanding of consumer attitudes towards organic food, since expectations may be higher and/or different as they are for conventional food.

The contribution of food chains to food quality may be crucial, since the processing, packaging and distribution may add or subtract quality to the organically produced agricultural raw materials or fresh produce.

That's why the concept of "care" should be extended to these phases and the "fork to farm" approach, including an effective and consistent supply management must be adopted. Technological and economic analysis of the organic food chain is also urgent because of the impact of a less efficient supply chain on the retail price. For example, since the extremely high prices for fresh organic products are probably the most important factor for a slowly growing demand in Italy, it is necessary to examine the economic aspects that influence such a situation.

In recent years the role played by several food scares has probably been of paramount importance to push the demand of organic food, despite the very high price premium. In 2003 market growth experienced its first slow down, especially in the large retailer sector, while the specialised retail (in countries where it is a relevant share of the market i.e., Germany and Italy) is facing a turbulent situation of restructuring. Presently, the increase of demand is not so exciting anymore (+1.4%), but it is still interesting compared to the food sector as a whole which is shrinking (–10%, Castellani, 2005). This is probably the consequence of a negative short-term phase for the Italian economy (many indicators showed a cycle of stagflation) that affected many industries and also large segments of consumers, thus reducing the available income for many households.

However, many operators were probably used to, or planned their projects counting on, a continuing period of growth. So a sort of depressive mood has recently pervaded the Italian organic food industry (Herrmann, 2005).

The perception of a state of crisis may be a chance to cope with the structural problems of the industry, and with the main aspects that are posing crucial questions to the industry operators, at any level of the supply chain.

A very important issue is that of "trust" in the system. This is of paramount importance at both the business-to-business and business-to-consumer levels. Certification of the product and assurance on the integrity of the system are hot questions.

The organic food industry is heavily-regulated and the organic supply chain is more fragmented than the conventional one. Since the premiums paid for organic food are mainly based on credence attributes, trust is a fundamental asset for the industry. The raising of concerns (Doward *et al.*, 2005) on the integrity of the organic food system may heavily damage all operators which are, in principle, self-interested in assuring organic quality (Jahn *et al.*, 2004 and 2005). Thus, compliance with the rules of organic production is a central issue since the beginning of the organic movement.

Trust creating operators play a central role, assuring both the supply chain actors and the consumers against possible disturbing levels of fraud or malpractice within the industry. In the organic supply chain, the creation of trust finds its formal application into a process and product certification, guaranteed by independent inspectors of the Organic Agriculture Certification Bodies (OACB). In Italy the number of authorised certification and control bodies is increasing with 19 such bodies currently, including societies working on the whole national territory and those operating just in the German-speaking province of Bolzano. Interesting research questions may be raised on this situation: is it just a burden for the low efficiency and the lack of clarity in the image towards the consumer, or is it also an opportunity for the differentiation in costs, control procedures (within the limits posed by compliance with law), and administrative burden, as well as in the meaning for the consumer.

The past development of the organic food industry has allowed a mechanism of self control or peer control to operate, but some researchers points out that this system may not be sufficient anymore to assure both the operators and the consumers against the risk of malpractices, hidden opportunistic behaviour and consumers' trust betrayal (Jahn et al., 2004).

Future studies could be aimed at analysing the role of OACB in creating trust between the operators involved into the organic food supply chain in Italy, to highlight the main aspects and problems of the organic quality assurance system, and the approaches adopted by the OACB in their field activities, considering their differences in structure, strategic objectives and degree of integration with the operators in the bottom and upper side of the supply chain.

4. CONSUMERS' ISSUES

On a consumer perspective some basic question are still hot in the Italian market.

- Is there room to expand the demand for organic food since potential demand still seems large even though actual demand does not grow accordingly?
- Is organic produce or an organic food product able to reach the organic consumer who is willing to pay the right price?

Zanoli (2004) highlights the fact that organic markets have different rates of growth, both at European and Italian levels. Actually, there are European markets that grew very rapidly in the past, but now seem to have reached the saturation level, while others are still growing and may expand their quota above the 2% average level of many markets, others that have remained very small so far, but may become emerging markets, according to a trend of economic development of the country economy (e.g. in Eastern Europe) or to an increasing popularity of organic food as a status and cultural (e.g. or Mediterranean Countries). Which socio-cultural explanation can be given for this? Can we relate this to a cultural specific belief or value system? To what extent are barriers for increasing the organic consumption more country specific than consumers' motives to buy organic? If we wish to develop the domestic market for organic produce, can we approach the consumers in European countries in the same way?

The north-south division among consumer countries and producer countries is still in place. While Germany is still the biggest market within Europe, Switzerland has the

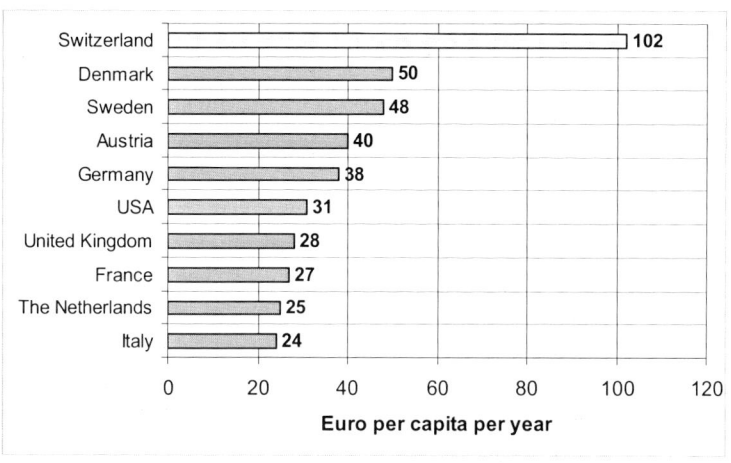

Figure 1. Per capita expenditure for organic food in Europe and US (year 2003), Source: FIBL, 2003

highest per-capita organic food expenditure (Figure 1). It is apparent that within and outside Europe there are differences in the success obtained by organic food. The consumption level in Italy has reached amounts not very different from those of France, The Netherlands, and the United Kingdom, although even within Italy a north-south consumer-producer split can be observed. The comparison of per capita expenditure may suggest that Italian consumers can still expand their expenses if the marketers will be able to develop a sound offer (Zanoli, 2004).

Yiridoe et al. (2005) provide a comprehensive evaluation of empirical studies comparing organic products and conventionally grown alternatives. Key organic consumer demand and marketing issues are considered, such as the most important attributes considered by shoppers when comparing organic with conventionally grown products, consumer knowledge and awareness about organic food, organic consumer attitudes and preferences, consumers' willingness-to-pay for organic products and related price premiums, as well as profile of the organic consumer.

The main problem is still that consumers are not consistent in their interpretation of what is organic, and most of them do not understand the details of organic farming practices and organic food quality attributes. In addition, a certain skepticism about organic labels may hold some consumers back from purchasing organic.

Consumers are sometimes motivated to buy organic food as insurance and/or investment in health, even if there are no scientific evidence of better health linked with organic food consumption.

On the price issue, it seems that the actual price is less important than the relative price level in respect with the conventionally grown product, and income elasticity of demand for organic foods is generally small.

Finally, organic fresh fruits and vegetables currently dominate the organic consumer's food basket.

In this framework, the Italian market brings some specificities, some of which are listed below:

- the Italian consumers have been met by the (actual and perceived) inflation effect brought by the introduction of the Euro currency more severely than in other European countries, and more than 60% of consumers state they did reduce their food basket expenses (Berardini *et al.*, 2006).
- the level of trust in Institutions is normally very low among Italian consumers, and this reflects also in the attitude of consumer towards the organic quality assurance system.
- a low level of awareness on the EU organic label and the absence of a national mark for organic products, together with a scarce information (even if it is increasing, according to Ismea, 2005) on the features of organic products do not help the consumer to recognize the organic product.
- very few national brands exist and hold a relevant market share in organic food.
- the role of large retail is less important than in other countries, even if increasing, and premium food are often bought in specialty stores, rather than in supermarkets; this reduces the impact of the entry of some large retail companies in the organic market, and also created some disappointment in companies that were hoping for a larger gain in market share.
- the interest of consumers is very differentiated across the country and is mainly concentrated in the northern areas, in opposition to the areas of production.

While organic food may be presented as a single value attribute to the consumer, it also may be added to existing higher value products, in order to reinforce and further differentiate it. On this regard, the issue of compatibility and synergies of the organic concept PGI/PDO local and traditional food specialties is particularly interesting, since some marketers argue they are not compatible, while other argue that for some niche product the "organic" image should be essential to create a real market for them.

As a matter of fact, even if the organic concept is in its essence a "global" one, it is often perceived as a signal of closeness between farming and consumption. Then joining the organic concept with a strongly local feature appears to be interesting.

An important issue may be also represented by the so called "low input" products (Zanoli, 2004), i.e., those products which, in the eyes of consumers, appear healthy and natural or have been produced in a way which makes them so. These products represent potential market substitutes for organic products (Midmore *et al.*, 2005). In Italy several retail chains have set up private labels aimed at giving the consumer assurance on a higher safety/quality of the product, leveraging on the Integrated Crop Management systems or similar production technologies. Moreover, for several years a potential "zero-residues" fruit product to be marketed has been under discussion.

The effects of the launch of such products may be controversial, on one side it may be beneficial since it widens the choice opportunities of consumers, on the other side it may increase the confusion and uncertainty about quality features of the different low input products offered, discouraging the consumers to buy both organic and other higher quality food.

According to a literature survey performed by Meier-Ploeger and Roeger (2004) appearance and taste are reported to be of importance for choosing organic food for Italian people. However, other studies indicate that Italian consumers do not seem to prioritise the appearance of products, thus indicating that they use other quality measurements apart from appearance when it comes to evaluating food. Health is important to the majority of the consumers, and this issue might be even more prominent than environmental issues in Italians' self-perception of their reasons for buying organic food. In general consumers buying organic foods seem to be more ethically concerned and idealistic than buyers of conventional food. The origin of the food is important, but one study indicates that consumers view the origin of the food as a proxy for quality. Animal welfare as an issue is absent of the reviewed studies. With regard to safety, worry is mainly expressed in respect to the use of chemical pesticides in agricultural production.

A reason of misunderstanding is the basicly process-oriented characteristic of organic certification. Since the Italian consumer is often more distrusting about public and private institutions compared to other European citizens, actual demand may be negatively affected causing it to be generally lower in comparison with the potential demand. The issue of trust needs to be addressed and analyzed more in depth in Italy.

5. POLICY AND TRADE ISSUES

Considering the issues just mentioned above and those discussed throughout this book, a critical issue is the reshaping of regulations regarding organic food in Europe.

In June 2004 the Commission adopted a European Action Plan on Organic Agriculture, aimed at promoting the development of both supply and demand of organic agriculture in Europe. It has involved several actions, some of which brought to amend the EC Regulation 2092/1991. However, several representatives still ask for changes. A draft of a new regulation should be discussed in June 2006, and a final version of the new regulation for organic agriculture should be issued in year 2009 (Piva, 2006).

The main issue is that the EU Regulation 2092/1991 covers a number of consumer perceptions, such as certification system, traceability, minimal use of additives, labeling concept and the use of organic raw material for food processing. However, other consumer expectations are not covered (and probably will not be covered). These other issues include careful and "organically sound" processing, quality and freshness, healthy nutrition or fair trade. These issues need to be addressed, either within the EU regulation or through joint initiatives of the organic operators.

5.1 Sustainable development

Relationships between organic agriculture and the policy for a sustainable development in less-favoured areas are becoming more visible and incorporated into policies. Provisions for farmers are linked increasingly to rural development plans instead of subsidies to production. It is necessary to find points of agreement between sustainability strategies and organic agriculture and food processing strategies[1]. EU recognizes the role of organic agriculture to foster environmental and sustainable development, but up to now the EU regulation has not included specific prescriptions on principles and

---

[1] The Italian Ministry of agriculture is funding the project SABIO "Sustainability of organic agriculture", co-ordinated by the INEA, National Institute of Agricultural Economics, aimed at assessing the profitability, assessing the value of externalities produced by organic agriculture, and at evaluating the policies for the Mediterranean organic agriculture (http://www.inea.it/sabio/).

guidelines for food processing. From an economic perspective, the introduction of new regulations should have an impact on processing technologies, as well as on labeling of organic products. The economic impact on cost and benefit as well as on the profitability of organic processing practices needs to be analyzed. In addition, consumer response to different labeling (e.g., not only "from organic agriculture", but "organic") may be different, and the analysis of the role of such labeling on consumer trust should be explored. It is crucial also because consumer concens about food quality appear to be connected to both food production and food processing.

The indirect influence of organic farming diffusion may also mean a shift of the convenience thresholds for adopting new technologies in light of sustainability for ecosystems, as far as they will be considered compatible or incompatible with the massive use of transgenic crops both in developed and in developing countries. For some authors, GM crops are not compatible with sustainable or organic farming or any other alternative form of production while others support an opposite view, claiming they would help to implement a better Integrated Crop Management (Ramanjaneyulu, 2006; Gould, 2006; Nassar, 2006; Malagoli, 2006). The menace to a potentially profitable alternative way of production represents an important opportunity cost to be computed into a cost-benefits balance for the introduction of GM crops [2]. Actually, this argument is a basic one for the Italian legislator's decision (supported by the majority of Italian public) to deny the admission of novel GM crops to open cultivation and to adopt a positive labelling of food containing ingredients derived from GM organisms for more than 0.9% of that ingredient.

5.2 Farmers' income

The issue of farmers' income is certainly a crucial one, since one of the main motivations for the production of organic food is necessarily based on the producer's satisfaction, which includes the realization of a satisfying income.

Several analyses have been made on the profitability of organic crops and farms (including those available in this book). Carillo *et al.*'s (2005) analysis of the economic results of the farms incuded in the panel of the RICA farm accounting system (including both conventional and organic farms) allowed a comparison of the organic part with the total panel, showing a different structure of farms, a lower productivity, but a higher profitability, mainly due to a lower cost of inputs.

However, the decrease in number of organic farms, mainly due to the lowered support from the common agricultural policy measures for environment protection and rural development, showed that participation in the organic movement was not economically sustainable without an external support for some of the operators.

Furthermore, there are large differences between industries and types of production, and a general advantage for organic farming cannot be stated, since it depends on many different factors (technical, entrepreneurial, market, etc.).

---

[2] The Italian Ministry of Education, University and Research is funding the project "Rural development, modern retail, food safety: organic farming perspectives in Italy" co-ordinated by the University of Neaples, aimed at making analyses on possible future competitive scenarios. Among other issues, the study will be focused on the impact of the introduction GM food on Italian markets and its effects on organic production costs when identity preservation and traceability costs are taken into account as differentiation strategy. (http://www.ricercaitaliana.it/prin/dettaglio_prin_en-2004079383.htm).

Moreover, some special conditions (such as, dispersion and fragmentation of production, lower level of association between producers, small quantities produced, and absence of a well developed organic supply chain or organic food processing facilites in many areas) create problems for the exploitation of the market potential for organic production in some local areas [3]. In these areas, the organic production must sometimes be sold as conventional production, losing the potential price premium.

Community supported agriculture (CSA) (entailing initiatives like box schemes and promotion and support for direct selling at farm's gate as well as collective purchasing groups) allow farmers to develop alternative distribution channels in local markets and to develop short chains. Thus CSA may be seen not only as a means of maintaining the cultural specificities of agriculture in local communities but also as a means for maintaining the income level of farmers at an acceptable level. It must be pointed out that these initiatives may be successful only when the local community is aware of the advantages of organic agriculture, not just on the side of food quality, but also regarding environmental and social impacts of growing organically. In this regards and from a research perspective, it may be interesting to use spatial models to analyse the influence of organic farms' geographical location/site variables on purchase intentions and willingness to pay of consumers, as well as on farmers' decisions to step into organic farming.

5.3 International trade

Although the amount of organic farmland continues to rise across the globe, most sales of organic food and drink are restricted to the industrialized world. North America and Western Europe account for roughly 97 percent of global revenues. Other important markets are in Japan and Australia (Zanoli, 2004). The consumer market lies in the industrialized countries (Europe and the USA in particular), whereas a major part of the production area lies in Oceania and Latin America [4].

Above and in other parts of this book the fact that in Italy most of the growth of organic farming is mainly driven by a policy support on supply was highlighted. On the contrary, the increase of interest in developing countries can be accounted for the supply-driven market growth of organic farming.

In developing countries, organic agriculture may be seen as an alternative to comply with complicated and demanding quality assurance schemes requested by the western markets. It is a possible way to overcome sanitary and phytosanitary (SPS) entry barriers

---

[3] The Italian Ministry of agriculture is funding the project "RISBIO: Research of Strategies to Support Organic Companies", co-ordinated by the Alma Mater Studiorum-University of Bologna, aimed at improving the competence of organic food companies, promoting the definition of production processes, and building an integrated supply-chain information system, in order to deliver a safer product to consumers (http://www.ambitalia.org.uk/Organic.htm).

[4] At the present state, there is a lack of reliable and easily obtainable market information on the production, demand, and trade of organic food. Basic data (number of certified organic holdings, land and cropping area) exists, but there is still no requirement to record or report other important economic data, that is not accessible for market information purposes. This is a problem for producers, processors and traders as they find more difficulties to identify opportunities within the market. An EU group has been created to address this issue, with the aim of building a framework for reporting credible organic production and market statistics. The EISFOM-project (European Information System for Organic Markets) will act as the dissemination and discussion point for projects involved in market intelligence (http://www.eisfom.org/). In Italy, a single National Information System on Organic Agriculture has been established (Decree 92721 del 23/12/2003) by the Ministry of Agriculture (www.sinab.it).

and access the rich western markets with a value-added product. For many producing countries the compliance with less restrictive rules, like EurepGAP good agricultural proctices, good manufacturing practices (GMP), or those of the European retail chains like British Retail Consortium or International Food Standards, may be similarly or even more difficult to achieve.

However, the problem of international trade of organic produce or food products cannot be reduced to a sleight of hand adopted in order to grant the products an easier access to markets that raise non-tariff entrance barriers. On this issue, future research is needed to understand whether the organic consumer in Italy is also interested in consuming imported organic food, whether it assigns the same *premium* price to imported organic food, whether it requires additional guarantees, etc.

Italy is a net exporting country, and most of the goods exported are sold in developed countries. The increasing demand of organic food products and the parallel decrease of organic farms in Italy raise the problem of supply of organic raw materials for processed food. The opening to international markets is, in a mid-term perspective, a need more than an option, but it asks for a set of different economic analyses. International trade may also be linked to other ethically driven certification requests, in particular, that of fair trade. Fair trade is a very important aspect that implies the consideration of internal coherence of organic production with the bundle of values that organic consumers may take into consideration.

6. FURTHER EMERGING ISSUES

Another interesting issue is the role of organic agriculture and organic food in the mega-trend called "globalisation". This trend runs opposite to the important issue in Europe, and particularly in Italy, of the value enhancement of local/traditional specialties and of the local markets. The organic movement is, in its native essence, a global movement, aiming to spread its practice all over the world. But the organic movement also has to cope with the issues of traditional processes, of food security in less developed countries, as well as with the problem of international trade. In some cases its target is the same of other quality food products, but its values may be perceived partly in contrast or may have synergic effects with other "soft" quality attributes, like traditional values, linkage with the territories, fair trade issues, etc. Further research in the Italian market is needed on these issues since Italian culture specificities and cross-cultural differences between the Italian, European as well the US situations must be taken into account.

In the future, the influence of an increased interest in ethnic cuisine, immigration and international tourism will probably boost the interest and create new demand for food products imported from abroad, expecially from outside the EU (Fischer, 2004).

In order to favour this process, improvement in the ability of organic production to ensure food security and provide sufficient yield is also needed. In light of an expansion of the quota of imported goods to cover the domestic demand of organic food, a relevant issue is the impact of a reduced land productivity often associated with organic agriculture. This is a sensible argument for some developing countries (e.g. China) that still face the challenge of providing enough food for their population and must accept a trade-off between a higher profitability of labour, lower environmental impact, better farmers, consumers health conditions, and a reduced availability of food.

In this regard, research on agronomic, crop protection, quality preservation issues as well as on marketing issues is important from an Italian perspective, but it is even more relevant for a further enlargement of the organic production to other areas of the world. Two projects regarding organic agriculture have been recently funded by the European Commission within the Asia-link Programme that are specifically intended to develop the human resources of higher education institutions in EU and Asia. The first one[5] is specifically focused on the development of organic agriculture in China. The second one[6] is focused on the marketing of quality, unique and organic food products. Both projects have an Italian co-ordinator. In both cases the interest in the Asian countries involved is very high.

Among the motivations of Asian producers, the interest for export market, particularly the need of overcoming SPS non-tariff barriers is certainly playing a relevant role. However, the development of domestic markets is starting up, and interesting experiences have been made involving the development of short distribution chains, community supported agriculture, and local development initiatives aimed at environment conservation and restoration.

Since the "organic concept" is gaining interest in larger groups of consumers both in Italy and abroad, further opportunities may be taken into consideration. In this respect, another new development may be represented by Natural, Organic and "Eco-Friendly" Pet Products. As far as we know there are no studies on this issue, but it is a matter of fact that pets are increasingly treated with a care similar to that reserved to a child in Italy, and the psychological pressure to assure them a better life, together with recent doubts on the animal feed industry may be powerful drivers for this market.

Organic agriculture is also interested in non-food crops, like flowers or textiles, that may receive a boost from a diffusion of the ethical consuming culture (e.g., see Perilli, 2006), since the adoption of a new life-style in food consumption may be reflected also on other aspects of an individual's life, like wearing environment-friendly clothes, using bio-engineering concepts and so on.

It is hard to say if this is a fad or a real trend, but it is surely worth to be explored.

7. REFERENCES

Berardini, L., Ciannavei, F., Marino, D., Spagnuolo, F., 2006, *Lo scenario dell'agricoltura biologica in Italia*, Working Paper SABIO no. 1, INEA—Istituto Nazionale di Economia Agraria, Roma.

Canali, G., 2005, Biologico: rilancio possibile in tre mosse, *L'Informatore agrario*, **32**: 29–30.

---

[5] Organic Farming: Social, Ethical, Economical, Scientific and Technical aspects in a global perspective (CN/Asia-Link/028), managed by the University of Turin—Centre of Competence for the Innovation in the agroenvironmental sector (AGROINNOVA) (IT); Università degli Studi di Tuscia (IT); Rheinische Friedrich-Wilhelms-Universität Bonn (DE); Wageningen University (NL); China Agricultural University (CN); Northeast Agricultural University (CN); Zhejiang University (CN); Qinghai College of Animal Husbandry and Veterinary Medicine (CN) (EuropeAid Cooperation Office, 2006, page 293).

[6] BEAN-QUORUM: Building an Euro-Asian Network for Quality, Organic, and Unique food Marketing (TH/Asia-Link/ 006) is managed by the Dipartimento di Economia e Ingegneria agrarie dell'Alma Mater Studorium—Università di Bologna (IT); Xinjiang Agricultural University (CN); Universität für Bodenkultur (AT); Thammasat University (TH); University of Gloucestershire (UK) (EuropeAid Cooperation Office, 2006, page 207).

Canavari, M., Centonze, R., Nigro, G., 2006, *Organic marketing and distribution in Europe*, Paper presented at the 2nd BEAN-QUORUM (TH/Asia-Link/006) seminar at the Thammasat University, Bangkok, March 1, 2006.
Canavari, M., Nocella, G., Pirazzoli, C., Regazzi, D., Rivaroli, S., Scarpa, R., 2005, Prospettive economiche di mercato della produzione ortofrutticola biologica: un'indagine sul gradimento dei consumatori, *Italus Hortus*, **12**(3): 27–37.
Canavari, M., Palmieri, M., Pirazzoli, C., 2005, Pesco con tecnica integrata e biologica: costi e redditività a confronto, *Italus Hortus*, **12**(3): 39–44.
Carillo, F., Doria, P., Marino, D., Scardera, A., 2005, Struttura e risultati economici delle aziende biologiche: un'analisi tipologica attraverso l'utilizzo della banca dati RICA, in: Cicia G., De Stefano F., Del Giudice T., Cembalo L., *L'agricoltura biologica fuori dalla nicchia: le nuove sfide*, Atti del 2° Workshop GRAB-IT sull'Agricoltura Biologica, Portici, 9 Maggio 2003, Edizioni Scientifiche Italiane, Napoli, pp. 119–131.
Castellani, N., 2005, Mercato bio: il passaparola non basta più, *L'Informatore agrario*, **32**: 34.
Chryssohoidis, G.M., Krystallis, A., 2005, Organic consumers' personal values research: Testing and validating the list of values (LOV) scale and implementing a value-based segmentation task, *Food Quality and Preference*, **16**(7): 585–599.
Cicia, G., Fersino, V., Marino, D., Schifani, G., Zanoli, R., 2005, Le frontiere della ricerca nel campo dell'economia del settore biologico, in: Cicia G., De Stefano F., Del Giudice T., Cembalo L., *L'agricoltura biologica fuori dalla nicchia: le nuove sfide*, Atti del 2° Workshop GRAB-IT sull'Agricoltura Biologica, Portici, 9 Maggio 2003, Edizioni Scientifiche Italiane, Napoli, pp. 23–68.
Doward J., Townsend M., Wander A., 2005, Britain's organic food scam exposed, *The Observer*, August 21; http://observer.guardian.co.uk/uk_news/story/0,6903,1553438,00.html, accessed on August 30, 2005.
EuropeAid Cooperation Office, 2006, *Asia-Link Programme Partnership Projects: factsheet compendium 2006 (draft)*, European Commission, Bruxelles, http://ec.europa.eu/comm/europeaid/projects/asia-link/asia-link_pfsc_155_draft.pdf accessed on June 1, 2006.
Fischer, C., 2004, The influence of immigration and international tourism on the demand for imported food products, *Acta Agriculturae Scandinavica, Section C—Food Economics*, **1**(1): 21–33.
Gould, C. (2006), GM crops *are* compatible with sustainable agriculture, *Science and Development Network*; http://www.scidev.net/Opinions/index.cfm?fuseaction=readopinions&itemid=468&language=1; Accessed June 8, 2006.
Herrmann, G.A., 2005, *Organic Market in Europe: Opportunities for Italy*, oral presentation during the SANA Bologna 2005 exhibition, Bologna, Sept. 8, 2005, Organic Services GmbH, München; http://www.organic-services.com.
Ismea/ACNielsen, 2005, *L'evoluzione del mercato delle produzioni biologiche: l'andamento dell'offerta, le problematiche della filiera e le dinamiche della domanda*, Ismea, Roma.
Jahn G., Schramm M., Spiller A., 2004, Institutioneller Wandel der Qualitätssicherung im ökologischen Landbau: Zur Selbstauflösung der Verbandskontrolle. Konferenzbeitrag zur 44. Jahrestagung der GEWISOLA, 27.-29.09.2004, Humboldt Universität Berlin.
Jahn G., Schramm M., Spiller A., 2005, The Reliability of Certification: Quality Labels as a Consumer Policy Tool, *Journal of Consumer Policy*, **28**(1): 53–73.
Meier-Ploeger M., Roeger, M., 2004, Comparison of consumer perceptions of organic food quality in Europe, In: Otto Schmid, Alex Beck and Ursula Kretzschmar (eds.), Underlying Principles in Organic and "Low-Input Food" Processing—Literature Survey, Research Institute of Organic Agriculture FiBL, Frick, Switzerland, pp. 39–44.
Malagoli, C., 2006, *Etica dell'alimentazione*, Aracne, Roma.
Midmore, P., Naspetti, S., Sherwood, A-M., Vairo, D., Wier, M., Zanoli, R., 2005, *Consumer attitudes to quality and safety of organic and low input foods: a review*, Integrated Project No. 506358 "EU-QLIF: Improving quality and safety and reduction of cost in the European organic and "low input" food supply chains", deliverable 1.2; http://www.qlif.org/research/sub1/QLIF_Review_Reanalysis_%200509.pdf, accessed on April 6, 2006.
Naspetti, S., Zanoli, R., 2005, L'analisi mezzi-fini: un'applicazione allo studio del comportamento del consumatore dei prodotti biologici, *Rivista di Economia Agraria*, **LX**(1): 9–38.
Nassar, N.M.A., 2006, Are genetically modified crops compatible with sustainable agriculture? *Genetics and Molecular Research*, **5**: 91.
Nori, L., 2005, "Casa comune" del biologico verso il traguardo, *L'Informatore agrario*, **32**: 31.
Perilli, B., Oggi sposi, equi e solidali anche il vestito è bio, *La Repubblica*, April 24; http://www.repubblica.it/2006/04/sezioni/cronaca/sposi-solidali/sposi-solidali/sposi-solidali.html, accessed on April 24, 2006.

Piva, F., 2006, I dubbi del biologico sul nuovo 2092, *L'Informatore agrario*, **2**: 13–14.
Ramanjaneyulu, G.V., 2006, GM crops are not the answer to pest control, *Science and Development Network*; http://www.scidev.net/Opinions/index.cfm?fuseaction=readopinions&itemid=467&language=1; Accessed June 8, 2006.
Yiridoe, E.K., Bonti-Ankomah, S., Martin, R.C., 2005, Comparison of consumer perceptions and preference toward organic versus conventionally produced foods: A review and update of the literature, *Renewable Agriculture and Food Systems*, **20**(4): 193–205.
Zanoli, R., Gambelli, D., Naspetti, S., 2003, Il posizionamento dei prodotti tipici e biologici di origine italiana: un'analisi su cinque Paesi, *Rivista di Economia Agraria*, **LVIII**(4): 477–510.
Zanoli, R., 2004, *How country-specific are consumer attitudes towards organic food?*, Paper presented at the IBL-meeting "Looking for a market! Which knowledge is needed for further development of the market on organic farming?" Wageningen, 23 November 2004; http://library.wur.nl/biola/bestanden/1740084-1.pdf, accessed on April 6, 2006.

CURRENT ISSUES IN ORGANIC FOOD: UNITED STATES

Kent D. Olson[*]

SUMMARY

In this chapter, we look at the current issues surrounding organic food in the United States. We first look at production issues such as profit, yield, crop insurance, genetically modified organisms (GMOs), input availability, market information, number and size of organic farms, and development pressures. We then look at distribution and marketing issues such as the definition of organic, market growth, supply chain infrastructure, integrity of organic products within the food distribution system, consumer trust, and potential demand for organic food. In section 4, we look at current policy and trade issues such as organic certification rules and standards, the need to educate consumers on the meaning of those rules and standards, the impacts of federal farm policy, subsidization of organic farming practices within federal farm policy and crop insurance programs, and global trade in organic products. In the last section, I briefly discuss the critical need for the future research focused on production and marketing problems and opportunities specific to organic food especially pest control, risk management tools for organic farmers, supply chain for organic products, consumers' understanding of the certified organic label, and the ability of organic production to meet food needs.

1. INTRODUCTION

In this book, authors have identified and analyzed several issues relating to organic food from both the consumers' and producers' viewpoints. In this chapter, I discuss other current issues for organic food in the U.S. that are not addressed in preceding chapters.

---

[*] Department of Applied Economics, University of Minnesota, St. Paul, Minnesota. Thanks to Paul Porter, University of Minnesota, for his review of an earlier version of this chapter, but all errors and omissions remain the responsibility of the author.

I have organized these issues into four areas: production; distribution and marketing; policy and trade; and future issues.

2. PRODUCTION ISSUES

In this section, I discuss several issues and obstacles faced by farmers: profit, yield, crop insurance, contamination from genetically modified organisms (GMOs), input availability, market information, number and size of organic farms, and development pressures near urban areas. Farmers also have concerns, issues, and obstacles related to distribution, marketing, policy, and trade. These issues are discussed in the following sections and not covered in detail in this section.

Questions surrounding profit and yield are among first raised by farmers considering the transition to organic farming. In general but not in all, the studies cited below report profits from organic farming to be equal or higher than conventional farming even though yields are found to be equal or lower. The profit picture is better than the yield picture because organic crops have lower production costs and higher prices compared to conventional farming. Using data from the Rodale Institute's Farm Systems Trial Pimental *et al.* (2005) report corn and soybean yields to be similar between the organic-animal, organic-legume, and conventional systems after the transition period. Other studies report organic yields (especially corn yields) lower than conventional yields (e.g., Porter *et al.*, 2003; Smolik *et al.*, 1995). In three geographically separate studies, profit levels for organic and conventional systems were estimated to be equal without organic premiums and higher for organic systems when historical organic premium levels were included (Hanson and Musser, 2003; Mahoney *et al.*, 2004; Smolik *et al.*, 1995). Other studies show profit to be higher for conventional systems (e.g., Dobbs and Smolik, 1996). Thus, the issue of yields and profit is still not clear and remains an obstacle for farmers considering the switch to organic. As noted in the last section of this chapter, an issue for the future is how to increase more research at the local level on organic production.

The debate about whether organic yields are equal to or lower than conventional yields is joined not just by farmers but also by many concerned with total food supply compared to nutritional needs in the world. Since organic certification requires a rotation different from conventional farming's alternating corn and soybean, a farm could not produce the same total amounts of corn and soybean even though the yields are the same on a per acre basis. The issue quickly involves international trade since the level of grain exports would be affected if total production decreased due to a combination of lower yields and a different rotation. This issue can also quickly involve international development as a response to the need to develop production in developing areas of the world as export potential and thus trade could decrease. Thus, questions of what else can be produced and what is the nutritional value (as well as monetary value) of the alternatives need to be answered. These questions are intertwined with the question of the impact of shifting acreages and production levels on market prices.

Another aspect of a farmer's decision to adopt organic production methods involves yield and price variability and the availability of crop insurance for organic producers. Current law states that crop insurance products should be developed for crops grown under organic management. These products are being developed, but in practicality, many obstacles remain for organic farmers to fully benefit from the crop insurance in the same way that conventional farmers do—even if policies have become available for specific organic crops. These obstacles include the price elections used for crop insurance being derived from conventional products and not reflecting the higher prices paid for organic products, the design of the policies (e.g., yield or revenue, coverage levels) and the lack of coverage of loss of sales and markets due to accidental contamination of crops from GMOs (Hanson *et al.*, 2003). The lack of understanding organic management can also hinder the definition and acceptance of what are good (and thus insurable) organic practices.

On a broader scale, organic farmers also fear the loss of market share if entire regions are considered contaminated by processors and their consumers. Even though organic certification is based on production methods and not on zero residues, entire lots and shipments from certified producers may be rejected due to measurable levels of GMO contamination. The problem of defining "GMO-free" has moved from an academic question to a practical measurement question of what is an acceptable level for processors.

Organic farmers also have concerns about the availability of organic inputs (Hanson *et al.*, 2003). For example, the supply of organic pesticides is apparently not increasing as fast as the demand for them. The current and future availability of crop varieties preferred by organic farmers and consumers is a concern among organic farmers. For organic livestock producers, the supply of organic feedstuffs is of major concern. This concern will likely be alleviated as the production of organic livestock and organic crop commodities (mainly corn and soybean) increases in a balanced fashion in the future. However, if the balance is not maintained or reached, concerns have been raised about the ability of some to weaken the definition by being allowed to waive the organic feed requirement if sufficient supplies are not available in the market. This is discussed in more detail in the section on policy.

Organic production may also be limited by the availability of credit as long as organic practices are not fully understood or accepted by creditors. This lack of credit will be a detriment to increasing production in total and in terms of who is producing in the future. While creditors who do not fully understand organic farming may be less inclined to extend credit, it is also true that creditors will look favorably on any producer (organic or conventional) who has a demonstrated ability to pay debts in a timely manner.

Organic farmers also express a need for better market information. The same level and quality of public information supplied for conventional markets is not available for organic markets. In the absence of public collection and reporting, private information suppliers have filled in the gap. Consequently, this market information for organic products is available for a fee, not free as it is for conventional products. Even after payment of the private fee, a producer may note that the quality of information is not as strong as the quality of conventional market information. This is likely due to three reasons. First, the conventional market is so large that information on traded prices and quantities are readily available in public places; the organic market is not as large and organic products are not traded as publicly as conventional products. Second, the organic market is often closer to the consumer, so organic price information is often held much closer by the trading parties and neither released to public agencies nor published in public media. Third, the federal government has a long history of collecting and publishing price information for conventional products. The nascent organic market does not have this tradition of working with the federal market agencies.

As noted by Greene (in this volume), the supply of organic food has and is increasing. This growth is a concern to producers if the demand for organic food does not keep up with the growth in supply and prices decrease in the organic marketplace. This growth comes from three sources: an increase in the number of producers, an increase in the average size of producers, and imports. Each of these sources creates its own set of concerns. Imports of organic products is discussed in the later section on trade, but let's consider the wider implications of the first two sources.

As their numbers increase, the population of organic farmers will become more diverse in their reasons for converting to organic production, in their size, and in their management objectives. This diversity will likely be both good and bad. Farmers who are converting for ethical reasons will likely follow the strict spirit of organic regulations; however, concerns are raised about those who are viewed as chasing the organic dollar in the marketplace and thus perceived as not being so judicious in their adherence to the organic regulations. This potential weakening in the standards is discussed again in the section on policy. On the good side of growth in the numbers is that more producers (whether converting for ethical or economic reasons) will mean more approaches to entering the organic supply chain, new ideas in marketing, and creation of new organic buyers in new markets.

The debate in conventional agriculture about the industrialization of farming or about factory farms, is also taking place with organic farms. For many traditional organic farmers and consumers, being organic means being small and operated by one farm family. However, as management knowledge increased and the opportunity of greater profits in organic products increased, current farmers have become larger and large farms converted (or are considering converting) to organic production. USDA's organic regulations do not limit the size of organic farms, and these larger farms certainly help meet the growing demand for organic food. However, the debate over size is a heated debate. Economically, the debate is the same for organic and conventional farms. As Gardner (2005) and others point out, small operations can be competitive. But the ability to manage larger operations and increase income per family in that way means that some farmers will increase the size of their operations—both conventional and organic farmers will grow. The concerns about farm size by organic farmers come in three ways: the competition from larger organic operations (just as in conventional farming); the potential damage to the consumers' perception of organic food being produced on small, family operated farms and thus their willingness to pay a premium for that food; and environmentally, the belief that a large farm may not be as good a steward of the land as a small producer.

Development pressures due to being close to cities is a major concern for many farmers. This issue is present for both conventional and organic farmers. On one hand, the farmers are benefiting with increased asset values and better retirement portfolios. Direct marketing of products can be easier and more profitable with consumers moving closer to production. On the other hand, they face increased traffic, complaints about odors, dust, and noise, and the potential inability of the next generation to continue farming on the same land. For organic producers who wish to continue producing, relocation comes at a greater cost compared to conventional growers due to the need to spend three years recertifying new ground and having to move away from land they have spent perhaps years developing and nurturing for organic production. While this relocation cost would affect all organic producers, organic fruit and vegetable producers are especially affected since they are more likely to be located close to urban areas and the establishment of fruit trees and bushes takes more than the three year certification process.

3. DISTRIBUTION AND MARKETING ISSUES

The issues involved in distribution and marketing of organic products start with the definition of organic and include market growth, supply chain infrastructure, integrity of organic products within the food distribution system, consumer trust, and potential demand for organic food. The concerns, issues, and obstacles related to policy and trade involving distribution and marketing are discussed in the following section and not covered in detail in this section.

Even though the USDA standards are specific, consumers still have many choices in terms of organic or natural foods. They have not become automatically and solely loyal to USDA's labels. Retailers have found consumers willing to buy other products promising to be safer for themselves and the environment. USDA standards have three variations: organic, certified organic, and 100% organic. Processors and retailers, in their efforts to increase market share, use the USDA definitions as well as terms such as "natural" and "environmentally safe," for example. These latter terms are not defined in any government regulation or industry standard. But with the consumer looking for food produced with fewer pesticides and antibiotics and with production practices safe for the environment and workers, retailers are willing to help them think they are finding safe foods. For the organic market, the issue involves determining what the consumer wants and how that affects the demand for USDA certified organic products.

Growth in the demand for organic food and how it balances with growth in the supply of organic food will determine how market premiums will change at both the retail and farm levels. In some situations, the local connections between growers and consumers (and the trust that develops between the parties) means that organic certification is not always sought by the grower nor by the consumer. However, with demand growth in larger supply chains, the demand for certified organic products will grow with the certification allowing trust in the quality of the food rather than the personal connection between grower and consumer. So the question for the distributor and retailer is the rate at which the demand for organic foods will increase in conventional retail stores and, thus, the demand for their supply chains to supply certified organic products.

Demand growth also depends on the strength of consumer trust in the organic supply chain (as well as on the organic producer). Trust is eroded if they lose confidence in accreditation bodies, rules, boards, market participants, and others in the system to supply the safe, wholesome food they want. As discussed in the next section, organic producers and certifying agencies see this threat of broken trust and so fight to make sure the USDA certification label attains and maintains a high trust among consumers. More instances of food safety threats (such as BSE) may erode consumer trust in USDA's ability to protect the conventional supply chain and increase the demand for certified organic food.

Market research shows an interesting paradox that could mean more increases in the demand for organic products. The paradox is in the difference between survey estimates of consumers' willingness-to-pay (WTP) of 20–30% more for organic products compared to the actual margins seen in the retail outlets of about 50%. If these consumers are indeed potential costumers just waiting for retail premiums to drop, the total market for organic products is much larger than current levels. Since the supply of organic products supports the 50% margin, the strength of premiums for producers may be quite strong.

Demand growth means need for adjustments and expansion in the supply chain infrastructure: stores, wholesalers, warehouses, information technology, people, trucks, and so on. Identity preservation of organic products (especially fresh fruits, vegetables, and meats) complicates and increases the cost of supply chain expansion for the organic market compared to increasing capacity in the existing chain for conventional products. Growing pains are bound to occur.

4. POLICY AND TRADE ISSUES

Policy and trade issues affecting organic food production and consumption include the continuing efforts to protect and avoid the erosion of organic certification rules and standards, the need to educate consumers on the meaning of those rules and standards, the impacts of federal farm policy, subsidization of organic farming, crop and revenue insurance for organic producers, the definition of good farming practices within federal farm policy and crop insurance programs, and global trade in organic products.

The protection of the USDA organic standards and regulations remain a prime concern for the National Organic Standards Board and the organic industry. Since many people and companies perceive it as more lucrative and with higher growth potential than the conventional food market, the organic market attracts many who would rather try to subvert or avoid having to meet regulations viewed as costly. Violations, requests for variances, and lack of enforcement erode the standards and potentially weaken consumer trust of the organic label and thus dampen demand for organic food.

This concern is not an idle concern. An example of the reason for the concern can be seen in an attempt to subvert the standards, not by trying to change the rules directly, but to take away USDA's ability to spend money to certify that the supply of certified organic feed was insufficient in the market. This example began with a procedure included in the regulations to allow the feeding of non-organic feed to certified organic livestock when the supply of certified organic feed is not sufficient. The regulations require the USDA to do a market evaluation to make this determination after a filing has been made for this exemption from the feed requirement. If the USDA determines that there is sufficient certified feed available the feeding of non-certified feed is not allowed. The subversion occurred when a local Congressman filed a bill to restrict USDA's budget for market evaluation on this issue. If the market evaluation could not be done, the reasoning apparently went, the USDA could not rule that there was sufficient certified feed, so the exemption could not be denied. When this attempt became public, the outcry caused the bill to be dropped.

Even though the standards have been in place since 2002, this is a relatively short time for many consumers to understand what the organic labels mean. They may also be confused by the presence of similar labels such as "natural" and "produced with environmentally sensitive practices." So, a future issue for organic industry is the education of consumers on the meaning of a the USDA certified organic label.

As this is written, one of the biggest questions is what the next U.S. farm bill will include. If farm income support is based on commodity production, the growth of organic farming practices will not grow as fast as if income support is allocated to farmers on a per farmer basis and farmers are given more flexibility to choose crops and production methods without fearing the loss of government payments. An increase in the support of conservation programs in the new farm bill will most likely increase the interest and adoption of organic methods. Since organic farmers are required to utilize a rotation that puts more of their land in non-program crops compared to a non-organic farmer, they are currently at an economic disadvantage. Any alleviation of this situation in the farm bill will increase adoption of organic practices, however, changing these rules have, in the eyes of this author, a low probability of occurring.

While the European Union directly subsidizes organic farms mainly for environmental and food safety concerns, the U.S. relies on indirect market support and not on direct support (Dimitri and Oberholtzer, 2005). The U.S. supports organic farming through defining organic standards. However, the direct support of organic farmers is being discussed in terms of its potential benefits to the environment. This interest has created action at the local level, not at the federal or state level. Recently, the Board of Supervisors in Woodbury County, Iowa, passed a proposal to offer farmers property tax rebates up to $50,000 if they go organic (Hytrek, 2005).

One of the main ways that U.S. farmers reduce risk is by buying crop insurance which is subsidized well by the government. Both yield and price can be covered by current policies. It is easily available for conventional farmers and usually required by lenders as one of the requirements for obtaining credit. New insurance products are available and being improved for livestock producers.

Organic farmers would like to improve their ability to utilize crop and revenue insurance as a risk management tool. Until recent years, insurance products were, in one sense available for organic producers but not widely used due to problems of defining good management practices and lack of coverage for many crops grown by organic farmers. Since organic practices were not accepted, in many instances, as good management (compared to accepted conventional practices), losses could not always be certified as losses due to factors other than management and thus insurable. So organic farmers were not effectively protected since they could not show they followed accepted cultural practices. While some policies have been developed, few organic farmers are participating partly due to the use of conventional prices, not organic prices, being used to value losses.

The definition of good practices is done by a local, county committee established by the USDA's Farm Service Agency (FSA). Since organic farmers are a very small minority and usually not part of the mainstream of farmers, they were not usually a member of these committees. Thus, their voice in defining good practices in an organic system was not been heard in earlier years. However, through the efforts of the organic industry, the 2002 farm bill included provisions to add insurance products for organic farmers. Thus, while not through the local committees, organic farmers are being able to help define good practices. The future issue is to improve the definition and to expand its geographical applicability.

Several international trade issues can affect organic farming. One recent concern is the importation of products that meet organic standards in the originating country that are less stringent than U.S. standards. The ability to import and sell products that do not meet U.S. organic standards raises the concern that the value of the USDA organic label will be diminished. This trade issue has not been addressed to the satisfaction of many in the traditional organic supply chain.

The impact of the recent WTO ruling that certain cotton support payments are trade distorting may cause changes in federal farm policy away from direct support of production to support of farmers as farmers and, thus, allow farmers more flexibility to choose crops and crop production methods. This flexibility will allow farmers to choose perhaps organic methods and rotations that allow organic certification. As organic certification becomes more accepted internationally, some trading partners may use that certification as part of the product standards for what they import. For those countries and companies desiring to avoid GMOs, organic certification may become a requirement that could essentially used as a trade sanction tool. However, this author again puts a low probability on this occurring.

5. FUTURE ISSUES

While there are many issues that one could list for the future, the following are what I think are the more critical issues at this point in time. One of the critical needs is for more academic and private research focused on production and marketing problems and opportunities specific to organic food. For many decades, research has followed the path of developing conventional agriculture and foods. For organic food to grow in market share, research needs to be done on the improvement of organic pest control, risk management tools for organic farmers, the supply chain for organic products, and consumers' understanding of what the certified organic label means. As the whole food supply chain becomes more international (including the organic food chain), issues surrounding quality and especially lack of consistency in organic standards need to be identified and settled.

Improvement in the ability of organic production to meet food needs is also needed. This comes in part in the improvement in organic yields but also in the change in diets to those that require less meat (and thus less grain production).

6. REFERENCES

Dimitri, C., and Oberholtzer, L., 2005, *Market-led versus government-facilitated growth: Development of the U.S. and EU organic agricultural sectors*, WRS-05-05, ERS, USDA.

Dobbs, T.L., and Smolik, J.D., 1996, Productivity and profitability of conventional and alternative farming systems: A long-term on-farm paired comparison, *Journal of Sustainable Agriculture*, 9(1): 63–77.

Gardner, B., 2005, The little guys are O.K., *New York Times*, Op-Ed Contributor; www.nytimes.com/2005/03/07/opinion/07gardner.html?th, accessed March 7, 2005.

Greene, C., 2006, An Overview of Organic Agriculture in the United States. In this book: Organic Food: Consumers' Choices and Farmers' Opportunities, New York: Kluwer Academic/Plenum Publishers.

Hanson, J.C., Dismukes, R., Chambers, W., Greene, C., and Kremen, A., 2003, *Risk and risk management in organic agriculture: View of organic farmers*, Working Paper 03–03, Department of Agricultural and Resource Economic, The University of Maryland, College Park, MD.

Hanson, J.C., and Musser, W.N., 2003, *An economic evaluation of an organic grain rotation with regards to profit and risk*, Working Paper 03–10, Department of Agricultural and Resource Economics, University of Maryland, College Park, MD.

Hytrek, N., 2005, County approves tax rebates to organic farmers. Sioux City Journal, published June 29, 2005, http://www.siouxcityjournal.com/articles/2005/06/29/news/local/423456a971cc9c068625702f00079c2d. txt, accessed December 12, 2005.

Mahoney, P.R., Olson, K.D., Porter, P.M., Huggins, D.R., Perillo, C.A., and Crookston, R.K., 2004, Profitability of Organic Cropping Systems in Southwestern Minnesota, *Renewable Agriculture and Food Systems*, **19**(1): 1–12 (Also reproduced in this book).

Pimental, D., Hepperly, P., Hanson, J., Douds, D., and Seidel, R., 2005, Environmental, energetic, and economic comparisons of organic and conventional farming systems, *Bioscience*, **55**(7): 57–582.

Porter, P.M., Huggins, D.R., Perillo, C.A., Quiring, S.R. and Crookston, R.K., 2003, Organic and other management strategies with two- and four-year crop rotations in Minnesota, *Agronomy Journal*, **95**: 233–244.

Smolik, J.D., Dobbs, T.L., Rickert, D.H., 1995, The relative sustainability of alternative, conventional, and reduced-till farming systems, *American Journal of Alternative Agriculture*, **16**: 25–35.

INDEX

Certification; v; 5; 10; 11; 18; 19; 20; 21; 26; 48; 55; 56; 59; 67; 68; 71; 76; 94; 97; 107; 116; 117; 118; 119; 122; 123; 126; 127; 128; 129; 134; 140; 143; 144; 154; 162; 174; 177; 180; 189; 192
   accreditation; 20; 189
   EC regulation 2092/1991; 4; 5; 9; 11; 94; 95; 100; 108; 128; 143; 177
   eco-labeling; 20
   logo; 5; 95
   National Organic Program; 5; 16; 28; 67; 68; 71; 76; 94; 111
   organic; vi; vii; 11; 18; 21; 22; 23; 25; 26; 48; 57; 61; 68; 71; 72; 80; 109; 125; 127; 128; 139; 141; 144; 145; 162; 177; 179; 185; 186; 187; 189; 190; 192
   quality; 107; 116; 117; 119; 122; 123
   standards; 17; 18; 20; 23; 27; 139; 190; 191; 192
Consumer
   awareness; 4; 110; 175; 176
   demand; vi; vii; 4; 11; 14; 17; 21; 25; 31; 32; 50; 51; 54; 58; 59; 60; 66; 72; 80; 98; 101; 103; 106; 108; 109; 110; 122; 123; 125; 126; 127; 128; 130; 131; 132; 134; 136; 137; 138; 139; 140; 141; 156; 157; 159; 165; 172; 173; 174; 175; 177; 179; 180; 182; 185; 187; 188; 189; 190
   loyalty; 63; 106; 107; 109
   readiness; 50

Consumer research
   choice model; 34
   contingent valuation; vii; 131; 141; 156
   experimental design; 116
   questionnaire; 11; 52; 93; 99; 125; 130; 144; 145; 150
   revealed preferences; 116; 124
   stated preferences; 116
   survey; vii; 4; 5; 7; 18; 28; 49; 51; 55; 59; 60; 61; 63; 93; 98; 99; 100; 102; 104; 108; 109; 115; 119; 122; 123; 125; 128; 130; 131; 134; 139; 143; 144; 145; 158; 162; 164; 167; 176; 189
   willingness to pay; 49; 117; 118; 119; 121; 122; 123; 125; 128; 131; 135; 136; 137; 138; 139; 140; 141; 143; 144; 145; 146; 148; 149; 150; 179; 188; 189
Conventional food; 6; 63; 173; 177; 190

Distribution
   direct selling; 54; 104; 105; 106; 107; 109; 110; 179
   export; 3; 7; 44; 98; 181; 186
   farmers' market; 158; 162; 164; 165; 166
   food service; 6; 51; 59; 104; 105; 106; 109
   franchising; 6; 56
   grocery; 21; 66; 119; 122; 157; 166
   international trade; 171; 180; 186; 191

Fair Trade; 162
large-scale retail; 3; 4; 6; 7; 49; 50; 54; 56; 59; 60; 62; 93; 98; 103; 104; 105; 106; 107; 109; 110; 130; 140
marketing channels; v; 59; 96; 104; 105; 108; 109
specialized retail; 6; 49; 50; 54; 56; 59; 62; 126; 144
supply-chain; 96; 99; 101; 104; 109; 179
wholesale; 56; 58; 59; 60; 96; 103; 104; 105; 106; 109; 160; 163; 190

Environment; v; vii; 4; 8; 11; 12; 14; 15; 16; 18; 19; 22; 24; 25; 26; 27; 32; 48; 51; 59; 63; 66; 80; 98; 106; 111; 115; 116; 117; 122; 123; 126; 128; 140; 148; 157; 158; 177; 178; 179; 180; 181; 189; 191
Externalities; 122; 177

Farm management
cropping system; 65; 66; 80
manure; 55; 61; 68; 70; 80
pesticides; 26; 28; 65; 66; 67; 76; 116; 161; 177; 187; 189
rotation; 22; 34; 61; 65; 66; 67; 73; 76; 81; 186; 191; 192; 193
weed control; 55; 61; 68; 70; 76; 80; 81
Farm results
income; 18; 42; 44; 102; 190
intermediate costs; 39; 40
production cost; vii; 48; 55; 56; 58; 60; 61; 62; 68; 70; 76; 80; 84; 85; 86; 87; 88; 89; 96; 102; 107; 109; 125; 128; 129; 130; 132; 133; 139; 145; 155; 172; 178; 186
profitability; vi; vii; 4; 12; 23; 24; 25; 27; 31; 32; 33; 34; 42; 44; 51; 54; 65; 66; 80; 83; 84; 87; 88; 90; 102; 106; 108; 110; 125; 127; 133; 140; 172; 177; 178; 180; 185; 186; 192
transportation cost; 8; 55
yield; 23; 25; 44; 62; 65; 66; 67; 68; 69; 72; 73; 75; 80; 85; 86; 87; 88; 90; 96; 98; 102; 108; 152; 172; 180; 185; 186; 191; 192
Farm structure
cattle; 11; 49; 50; 52; 53; 54; 55; 56; 57; 58; 59; 61; 62; 63; 124; 155
human resource; 100; 181
livestock; 4; 8; 11; 12; 19; 20; 25; 26; 48; 61; 62; 80; 164; 187; 190; 191
Food
fish; vii; 125; 126; 127; 128; 129; 130; 131; 132; 133; 134; 135; 137; 138; 139; 140; 142; 159
fruit; vi; 5; 7; 8; 17; 22; 25; 37; 40; 83; 84; 85; 87; 89; 116; 123; 124; 164; 176; 188
apples; vii; 88; 141
peaches; 86; 88
meat; vi; 6; 25; 47; 48; 49; 50; 51; 52; 54; 55; 56; 57; 58; 59; 60; 62; 63; 126; 143; 144; 145; 146; 155; 159; 161; 192
beef; vii; 8; 25; 33; 48; 49; 50; 51; 54; 56; 57; 59; 62; 143; 144; 145; 146; 147; 148; 149; 150; 151; 152; 153; 154; 155
poultry; 11; 14; 49; 51; 52; 53; 54; 55; 56; 57; 58; 59; 60; 61; 62; 159; 161
sheep; 8; 11; 25; 51; 52; 53; 54; 55; 56; 57; 58; 59; 61
Food safety; v; 50; 66; 162; 178; 189; 191
bovine spongiform encephalopathy; 4; 48; 63; 126; 144; 152; 164; 189

Genetically modified organisms; 4; 128; 172; 182; 185; 186; 187
Globalisation; 171; 180

Health; v; 4; 19; 20; 21; 27; 53; 55; 94; 98; 120; 126; 128; 129; 145; 157; 158; 161; 164; 175; 180

Integrated crop management; 85; 86; 87; 88; 115; 117; 119; 121; 122; 123

Managerial skills; 108

INDEX

Marketing strategies; vii; 47; 58; 71; 93; 98; 99; 101; 106; 107; 108; 109; 110; 116; 163

Natural foods; 21; 189

Organic farming; v; vi; 4; 8; 11; 14; 17; 18; 19; 20; 21; 22; 23; 24; 25; 26; 27; 31; 32; 33; 34; 36; 38; 39; 44; 48; 66; 80; 88; 95; 100; 106; 128; 129; 132; 139; 140; 171; 172; 175; 178; 179; 183; 185; 186; 187; 190; 191
Organic food; v; vi; vii; ix; 3; 4; 5; 6; 10; 16; 18; 21; 25; 49; 50; 59; 63; 67; 115; 123; 124; 126; 127; 128; 130; 134; 136; 137; 139; 140; 157; 158; 159; 160; 161; 162; 163; 164; 165; 166; 171; 172; 173; 174; 175; 176; 177; 178; 179; 180; 181; 182; 183; 185; 188; 189; 190; 192
Organic producers; 6; 8; 9; 11; 12; 13; 17; 20; 21; 22; 23; 26; 27; 28; 39; 49; 50; 93; 98; 104; 110; 143; 154; 163; 164; 172; 185; 186; 187; 188; 189; 190; 191; 192; 193

Policy; 26; 107; 109; 111; 116; 118; 122; 126; 171; 177; 178; 179; 185; 186; 187; 188; 189; 190

EU
  EC regulation 2078/1992; 8; 48
  farm; 185; 190; 192
  public support; v; 8; 14; 18; 26; 48; 110; 140
Premium price; vi; 4; 7; 8; 21; 23; 49; 50; 58; 62; 63; 65; 66; 68; 69; 71; 72; 76; 78; 79; 80; 81; 104; 108; 122; 127; 128; 131; 133; 135; 137; 139; 140; 146; 159; 166; 173; 175; 179; 180; 186

Risk; vii; 4; 21; 22; 23; 27; 56; 66; 72; 73; 78; 80; 103; 106; 107; 148; 172; 174; 185; 191; 192

Scenario; 11; 16; 62; 71; 72; 103; 111; 130; 181
Stochastic dominance; 72; 73; 80
Sustainability; vi; 17; 20; 22; 27; 28; 54; 66; 133; 162; 171; 177; 178; 182; 193

Transition; 8; 18; 22; 67; 71; 73; 186

Wine; vi; 7; 8; 52; 93; 94; 95; 96; 97; 98; 99; 100; 101; 102; 103; 104; 105; 106; 107; 108; 109; 110; 165

Printed in the USA